The Origins of
The International Vaccine Institute

A Personal History

Seung-il Shin

Author

Seung-il Shin

Former Senior Health Advisor and
Director of the IVI Project
United Nations Development Programme

———

Shin received undergraduate and graduate degrees from Brandeis University, and trained in human genetics at the University of Leiden. He was one of the founding members of the Basel Institute for Immunology in Switzerland. He then served as a Professor of Genetics at Albert Einstein College of Medicine in New York for 14 years, where his research was in somatic cell genetics and tumor biology.

He was a co-founder and CEO of Eugene Tech International, a biotech venture based in New Jersey. He led the development and commercialization of a low-cost hepatitis B vaccine in a joint effort with the New York Blood Center, and played a leading role in an international campaign to bring the new vaccine to the developing world.

The United Nations Development Programme in 1992 invited Shin to develop and implement a plan for an international center of vaccine sciences focused on the needs of developing countries under the aegis of the Children's Vaccine Initiative, a global movement to expand the availability of affordable vaccines for the world's children. The UNDP-initiative eventually led to the birth of the International Vaccine Institute (IVI) in 1997. In this memoir, Shin recounts how, despite some initial difficulties, international partners came together to create IVI and to find a home for it in Seoul, Korea.

The Origins of
The International Vaccine Institute

Seung-il Shin

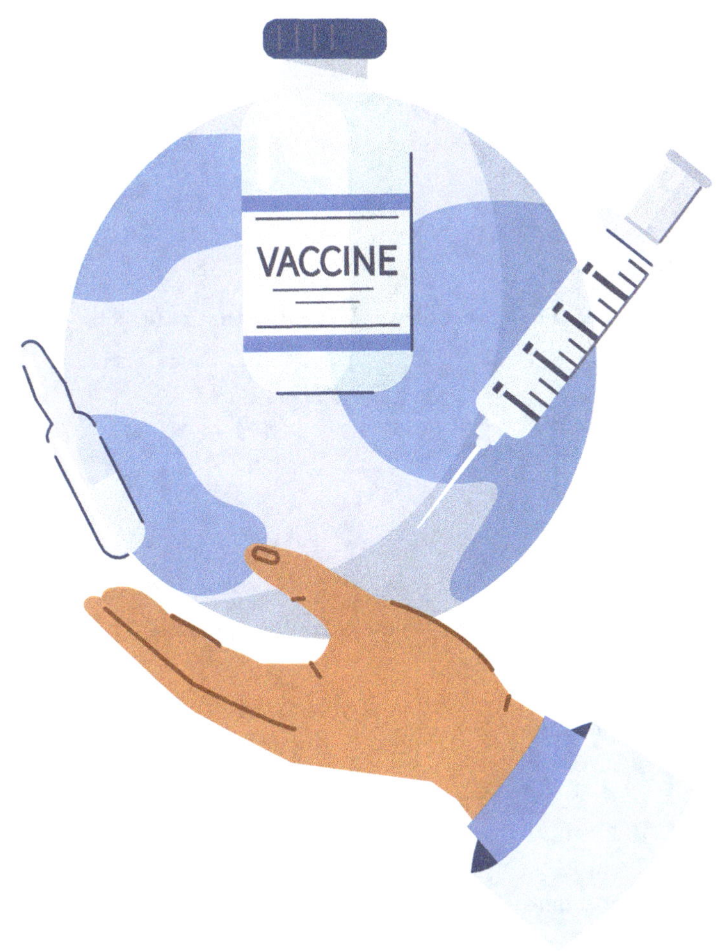

**The Origins of The International Vaccine Institute:
A Personal History**

Copyright © 2023 Seung-il Shin
ISBN 979-11-978605-2-2 (93470)

Edited by Park So-yeon
Designed by VUE

First published in 2023
Published by Goggas
Anyang-si, Gyeonggi-do, Korea
phone +82 70 8019 0433
goggasworld@gmail.com
www.goggasworld.com

Contents

Foreword

This is a highly personal account of the genesis of the International Vaccine Institute. Former colleagues of IVI and others who were associated with the founding of the Institute have urged me to leave a record of the key events that led to the Institute's establishment by the United Nations Development Programme (UNDP).

This memoir retraces the events of the early years of IVI, beginning with the launch in 1992 of a feasibility study for an international institute for vaccine sciences in Asia, and ending with the formal opening of the Institute in 1997. The Institute's evolution and development after 1997 are not covered here since they are available in the public domain.

I will first describe in some detail how the idea for an international vaccine institute first took shape, and then recount the early milestones at UNDP as the idea solidified into a real-life institution. Much of these events date back more than 30 years now, and therefore some of the details are subject to inaccuracies of personal memories. I hope that this very personal story will nevertheless serve as a useful guide to those interested in the origins and the early history of the institute.

1

Hepatitis B Vaccine and the Birth of an Idea

The idea for an international center for vaccine research, training and allied studies focused on the needs of the developing world began to take shape around 1990. It was directly related to my (though ultimately mostly futile) efforts to bring to the poorer countries of the world a new hepatitis B vaccine that we had developed in Korea based on the seminal work of Alfred Prince.

The hepatitis B virus (HBV) was discovered by Baruch Blumberg in 1967. The global disease burden caused by HBV was extremely high, and the virus was widely recognized to be the major cause of liver cancer, one of the most common and lethal cancers in many countries of the world but especially in developing countries. Dr. Blumberg's discovery won him the

Nobel Prize in 1976.

Based on the Blumberg discovery, both Merck Sharpe and Dohme of the U.S. and Pasteur Merieux of France each developed a vaccine against HBV by pooling the viral surface antigen from the blood plasma of virus careers and inactivating it using different methods. Independently, Alfred Prince of the New York Blood Center developed a more efficient process of inactivating the viral antigen by heating it briefly under controlled conditions. The Prince process also caused the antigens to coagulate, leading to increased immunogenicity, and opened the way for an effective but inexpensive vaccine against HBV. Vaccination of children against HBV, even in the poor countries, became a realistic possibility.

In 1984, Cheil Sugar & Co. of Korea and its U.S. subsidiary company Eugene Tech International (ETI) secured exclusive world-wide license to use the Prince method from the New York Blood Center, and successfully developed a large-scale process to produce a plasma-derived hepatitis B vaccine. The new vaccine was approved by the Korean FDA in 1985. Cheil Sugar & Co. was a part of Samsung Group company at that time, but later became an independent corporation with a new name, CJ. Since then CJ has itself become one of the largest Korean corporations with business activities in many fields, but perhaps best known outside Korea for its global entertainment business in music, cinema and other so-called K-wave culture products.

In 1984, I had taken a two-year leave from my faculty position as Professor of Genetics at Albert Einstein College of Medicine in order to lead the new biotech company, ETI. The company was founded in 1982 by six Korean scientists, including myself, who were senior members of universities and research laboratories in the New York metropolitan area. The founding members were Prof. Tong H. Joh (chemistry of neurotransmitters; Cornell University Medical Center, New York, NY); Prof. Yoon Berm Kim (a pioneer of using germ-free piglets for immunology research; Sloan Kettering Institute of Cancer Research, Rye, NY); Dr. Young Tai Kim (immunology; Rockefeller University, New York, NY); Dr. Hong Mo Moon (molecular biology; Roche Institute of Molecular Biology, Nutley, NJ); Dr. Kwang Soo Kim (biochemistry of medical diagnostics; New York State Institute of Health, Staten Island, NY) and Prof. Seung-il Shin (cell biology and genetics; Albert Einstein College of Medicine, Bronx, NY).

The founders of ETI wished to contribute to the development of bioindustry in Korea by training young Korean scientists in the newly emerging fields of genetic engineering and biotechnology. But ETI in reality remained a paper company because it did not have any funding or actual laboratory programs. In 1983, we agreed to make ETI a wholly-owned subsidiary of Cheil Sugar Co. of Korea, as Cheil and Samsung were looking for ways to enter the global biotechnology business. ETI would become the beachhead in America for the Samsung Group in genetic engineering and pharmaceuticals businesses. With Samsung's investment, ETI/Cheil opened a fully-equipped new laboratory in Allendale, NJ. Fresh graduates from Korean universities and graduate schools were brought to ETI to train under U.S.-based senior scientists.

When I was asked to lead the new ETI/Cheil company, while other ETI co-founders stayed with their institutions, my initial plan was to take a two-year leave from Einstein. But for me, as an experimental scientist, taking a long leave was in fact the same as resigning from my faculty position permanently. I had to let my graduate and postdoctoral students go and ask my laboratory staff, including two lab technicians and my secretary, to find new jobs. When I arrived at Einstein in 1972 from the Basel Institute for Immunology in Switzerland, I introduced the first athymic nude mouse colony for research use in an academic laboratory in the U.S. I could not continue that operation either. Leaving the Einstein faculty position for a corporate job was a major career-changing decision for me.

What finally prompted me to take the risky plunge was the promise of advancing Korean bioscience and introducing the new biotechnology industry that was just emerging in the U.S.

The hepatitis B vaccine was one of the first major commercial projects undertaken jointly by Cheil and ETI. The new vaccine we developed was an important breakthrough for global HBV control programs, because it could be produced in large quantities at low cost. In comparison, the first two plasma-derived HBV vaccines licensed by Merck and Pasteur Merieux were being marketed internationally at about $30-$50 per dose. Since three doses were needed to fully immunize a person, the Merck and Pasteur vaccines were too expensive for public vaccination programs for children in developing countries. From the very beginning, our goal at Cheil was to make the vaccine available for developing country markets for $1-3 per dose. A year earlier, Korea Green Cross company had received the Korean FDA approval for their own plasma-derived HBV vaccine, but the Green Cross vaccine was based on a process nearly identical to the Merck vaccine technology, and was therefore similarly expensive. Moreover, Merck prevented international sale of the Green Cross vaccine claiming that it violated Merck's patent.

Because of the technical breakthrough of the Prince

method, Cheil was in a unique position to make its HBV vaccine available to the developing world at an affordable price. The Chairman of the Samsung Group, B.C. Lee, wrote an open letter to the Board Chairman of the New York Blood Center (who was also the Chairman of the Citicorp.) stating that Samsung was prepared to donate 1 million doses of the Cheil vaccine for children in the neediest countries, and that Samsung/Cheil hoped to offer the vaccine for $1 per dose for public vaccination programs in developing countries.

In October 1984, while a crash program to develop the vaccine was still on-going at Cheil's laboratory in Korea, I organized an international conference on hepatitis B immunization for developing countries in Seoul with the help of Fred Prince and James Maynard. The goal was to explore ways to introduce international HBV vaccination programs when the low-cost Cheil vaccine became available. In addition to key players from the U.S. and Korea, senior scientists from Australia, India, Indonesia, the Philippines, Taiwan, Kenya and Nigeria also attended the conference. Prof. Xu Zhi-yi of Shanghai First Medical School, a renowned epidemiologist who carried out a clinical trial of our vaccine in Shanghai, was invited but could not attend, because the barriers of the Cold War era were still in place and Chinese scientists could not travel to South Korea in 1984. The conference generated tremendous enthusiasm and optimism for HBV vaccination programs in developing countries.

The optimism evident at the Korean conference led to the creation of *The International Task Force for Hepatitis B Immunization* at a follow-up meeting at Dr. Prince's office in New York in early 1985. The Task Force was composed initially of Alfred Prince, James Maynard, Richard Mahoney and Ian Gust. I was one of the originators of the Task Force, but did not join the Task Force because I was representing the vaccine producer company. The Hepatitis B Task Force regarded the inexpensive Cheil vaccine as the leading edge for a global campaign against hepatitis B, and actively promoted national HBV vaccination programs in developing countries in Asia and Africa. The high-level activities of the Hepatitis B Task Force also served to confer international credibility to the Cheil vaccine. The history and work of the HBV Task Force have been described in detail in a well-researched book by William Muraskin, *"The War Against Hepatitis B: A History of the International Task Force on Hepatitis B Immunization"* (University of Pennsylvania Press, 1995).

The Cheil vaccine was soon incorporated into the mandatory national immunization program for all newborns and young children in Korea. Korea was in fact the first major country to implement such a national program, and the Cheil vaccine was a major driver that made it possible. In 1987, Cheil made its vaccine available for $3 per dose to developing country markets, and announced that its aim was to eventually lower it to $1 for publicly funded international programs. Public health officials and vaccine companies from a number of countries came to meet with us in Korea to discuss introducing the Cheil vaccine into their national programs.

In 1986, I resigned permanently from the faculty of Einstein College of Medicine, and assumed the additional title of Executive Senior Vice President and Director of the R&D Laboratories of Cheil Sugar Co. with office in Korea, while continuing as the CEO of ETI in New Jersey. I began to devote much of my time to lead Cheil's efforts to take the hepatitis B vaccine to developing countries. For members of the International Task Force and me, it became an international public health crusade. However, for Cheil Sugar & Co., the hepatitis B vaccine project had to succeed as a business venture first. It was thus inevitable that I would face strong opposition from the senior executives within Cheil Co., whose main concern, understandably, was the profitability of the product. Only strong personal support from the company's CEO Young-Hee Sohn and the Head of the Pharmaceutical Business

Division Hyo-Gyu Kim allowed the crusade to continue. These two men saw the hepatitis B vaccine program as a long-term investment for the future, as well as a meaningful humanitarian program for developing countries that justified the investment.

During the period of 1988 to 1990, with the marketing staff of Cheil Co., I met with government officials and importing agents from China, Hong Kong, Taiwan, Malaysia, Singapore, the Philippines, India, Indonesia, Iran, Mexico, Uzbekistan and South Africa, as well as the former Soviet bloc countries of Russia, Bulgaria, Romania, Poland and Hungary. They came to Korea, but I also personally visited them. People from these countries were eager to initiate a national HBV vaccination programs with the new affordable vaccine. India was the first major country that granted the import permit for the Cheil vaccine, because the process was managed by an experienced private company, the Bharat Serums and Vaccines of Mumbai. The Indian effort was led by Prof. Vinod Daftary, a respected microbiologist who had participated in the first HBV vaccine conference in Seoul in 1984. The Philippines and South Africa then followed.

But none of the other countries actually managed to launch a national vaccination program. The major hurdle was securing the budget for such a program from their national governments.

Another hurdle was the registration of the new vaccine in each country, which proved to be a time-consuming and often highly bureaucratic process. High-level officials in 3 Asian countries indicated to me that they could not license a vaccine that was developed, produced and licensed in South Korea, which was generally regarded as just another Asian developing country. The Minister of Health of one Asian country told me privately that her government did not have the capability to properly evaluate a new vaccine, and therefore, in order to get an approval in her country, Cheil should submit additional data to show the vaccine's safety and efficacy, preferably based on clinical trials conducted in an "English-speaking Western country" such as the U.S. or the U.K. If any of these Western countries approved the vaccine, then her country will also approve it, she said. Given the high cost and long time required for such trials in the U.S., coupled with the limited market potential there, that was not a realistic option for us.

In addition to simple export agreements, Cheil was also willing to enter into technology transfer partnerships for local production of the vaccine. Several countries sought such an option, and we had reached advanced discussions with Mexico, Iran, Uzbekistan and Russia. With my Cheil staff, I visited the proposed sites for a vaccine production facility at two separate locations near Mexico City in 1989, and one proposed site at the State Vaccine Institute near Moscow in 1990. Despite multiple meetings and mutual visits with each country, however, none of the production partnerships materialized. Even though the proposed partners in these countries often included very senior officials such as the former cabinet secretary of health or the director of a government vaccine company, they failed to secure the needed investment capital or to assemble a group of engineers, production experts and managers necessary for such a major project.

Clearly, even for a vaccine that is essential for national vaccination goals, a developing country cannot effectively bring it into its national programs just because the vaccine is available at a low cost, either from a foreign producer or through local production. The country must also have its own regulatory capability to evaluate and license a new vaccine made by a foreign producer, or be able to assemble the multiple resources to build and manage a production facility.

After nearly three years of hard work that produced few visible successes, I came to the conclusion that, in the long run, affordable vaccines made available by foreign donors will not solve the long-term dependency of developing countries on outside donors. Since most vaccines are usually the proprietary products of the market-driven vaccine industry of advanced countries, developing countries will need to create regional or international mechanisms that will focus on their unique and specific needs. One way to achieve "vaccine independence" for developing countries would be the creation of an international, not-for-profit, cooperative organization focused on the needs of developing countries. Ideally, the organization should be located in a developing country, and staffed by scientists and supporting staff from developing countries as much as possible.

My initial working name for the organization was "International Center for Vaccine Research (ICVR)".

2

Enter Children's Vaccine
Initiative and UNDP

The Children's Vaccine Initiative (CVI) held its first consultative conference at WHO in Geneva in December 1991. CVI was an ambitious new global movement to accelerate the development and global supply of children's vaccines to protect the children of the world from vaccine-preventable diseases, especially in developing countries. The meeting was convened by the five co-sponsors of CVI, comprised of four UN agencies (UNDP, UNICEF, WHO, World Bank) and the Rockefeller Foundation. Dr. Philip Russell, former Director of Walter Reed Army Institute of Research, was Special Advisor to CVI.

I was invited to the meeting because Cheil was potentially a major supplier of affordable vaccines for the children of developing countries. Many senior scientists with whom I had worked closely for the hepatitis B vaccine movement also attended the meeting. The discussions at the two-day conference revolved mostly around the usual issues on promoting vaccine development, increasing vaccine availability, supporting national immunization programs for children in developing countries, and urging more international investments for these programs. It seemed to me that the discussions were almost entirely from the perspectives of the industrialized world institutions and large commercial vaccine companies. Looking around the meeting room, I was acutely aware that there were few representatives from developing countries who were personally engaged in vaccine R&D and vaccine production operations in the poor countries, which the CVI movement aimed to help.

I noticed that Frank Hartvelt of UNDP was the exception. He discussed the need for capacity building in developing countries related to vaccine development and production. During a recess, I spoke to Hartvelt about my observations of the day's proceedings, and told him that addressing vaccine-related issues of developing countries would benefit from more active participation by people from developing countries. He fully shared the same concerns and perspectives with me.

I told him briefly about my recent experiences with

the hepatitis B vaccine project, and said UNDP's extensive experience in capacity building in developing countries could be directed to a new international center for vaccine sciences within the context of CVI. Frank was very interested in the idea, and invited me to visit him in his New York City office for further discussion. A few weeks later, in early 1992, I visited Frank at his office at UNDP in the UN complex in Manhattan. We discussed the potential merits of a new international center for vaccine-related sciences focused on the needs of developing countries, to be located in a developing country and staffed as much as possible by scientists and managers from developing countries.

Frank Hartvelt was Deputy Director of the Division for Global and Interregional Programmes (DGIP) of UNDP. As the main administrative and programmatic arm of the United Nations, UNDP maintained country offices representing the UN in almost every developing country of the world, including South Korea. DGIP was the lead office of UNDP in capacity building programs, such as water and environmental management, climate change, public health resources and so on. It was therefore no surprise that Frank represented UNDP at the CVI Governing Council, and also that he was receptive to the idea of an international institute for vaccines for the developing world. At the end of our meeting, Frank asked me to write up my ideas in the form of a proposal to UNDP.

In April, I sent him a position paper in which I proposed a new international institute, to be established under UNDP leadership as a contribution to the CVI coalition. I changed the name of the institute from International Center for Vaccine Research (ICVR) to International Vaccine Institute (IVI), because the name ICVR seemed to be too narrowly focused on the R&D side. To respond to the much broader goals of CVI, I felt that the new institute should also deal with production and regulatory issues, as well as with providing assistance for national vaccination programs in individual countries.

In the proposal, I summarized the rationale for locating IVI in a developing country, and the need to mobilize the technical and managerial expertise that exists in many developing countries. I suggested that the emerging countries in East Asia and the Western Pacific region would be the best place to find a host country for an organization such as IVI. It should be possible, I argued, to persuade a developing country in this region to finance a significant part of the money to establish and support IVI, and to become an active contributing member to the global CVI movement.

The key parts of the proposal to UNDP are reproduced

below in its original form, as submitted to UNDP in 1992. The proposal was later incorporated *in toto* in the project document sent to prospective host countries of IVI issued by UNDP in 1993.

Proposal for an International Vaccine Institute:
A UNDP Initiative for CVI

By Seung-il Shin
April 1992

A. Background

Despite major strides in our ability to prevent and treat infectious diseases, they continue to be major causes of morbidity and mortality in both industrialized and developing countries. Many of these diseases take their greatest toll among children under 5 years of age: acute respiratory infections, diarrheal diseases and other potentially vaccine-preventable diseases account for over 80% of mortality occurring in this age group worldwide.

While a variety of preventive and therapeutic measures exist for reducing the impact of most infectious diseases, vaccines often prove to be among the most cost-effective means of disease control. This is especially so when the vaccines are also inexpensive to purchase and deliver. In certain circumstances, their application can even permit the global eradication of a disease, as has been achieved with smallpox and is being attempted with poliomyelitis.

Research and development of high quality, low cost vaccines to reduce the burden of disease, therefore, represents a global priority for social investment. It was in this context that the Children's Vaccine Initiative (CVI) was founded in 1991, to promote the development and stable supply of safe, effective and affordable vaccines against the major infectious diseases for the children of the world.

International efforts for the development of new and improved vaccines and for further expansion in vaccination coverage have gained a critical impetus and a new visibility through the creation and implementation of targeted programs under the CVI. Indeed, the CVI movement already represents a global groundswell of potentially revolutionary consequences, especially for the developing countries.

Recent assessments by the CVI have, however, revealed significant deficiencies in the quality and in global supply systems of vaccines used in children's immunization programs. Many governments fail to rigorously apply national and international standards for quality assurance. The absence of international regulatory mechanisms that are appropriate and accessible for vaccines produced in developing countries, and the difficulties in obtaining the latest vaccine production technology at an acceptable cost, have prevented many developing country producers from contributing more actively to the world's vaccine supply pipeline. In fact, these obstacles present major challenges not only to the ultimate success of the CVI, but also to the continued effectiveness of the highly successful Expanded Program on Immunization (EPI), which has done so much to immunize the world's children against major infectious diseases.

B. Why an International Vaccine Institute now?

On the other hand, recent dramatic advances in molecular biology, immunology and drug delivery systems offer an unparalleled opportunity for the development of new and improved vaccines. Most of these advances are made in small start-up biotechnology companies, academic laboratories or pharmaceutical research centers in a few advanced countries, that are not necessarily interested, nor can afford to be interested, in the general goals of public sector initiatives of global importance. Moreover, the relatively unattractive profit potential of children's vaccines in general, the liability exposure and other economic issues, make it unlikely that large multinational commercial vaccine producers will allocate major resources to turn someone else's nascent proprietary technologies into meaningful applications towards the development of "ideal children's vaccines" envisioned by the CVI, which are primarily directed towards the needs of the developing countries.

A not-for-profit international vaccine institute, established and governed by an autonomous international board, will be better able to serve a facilitating role in channelling appropriate technological innovations towards new vaccine production know-how.

Vaccine-related technologies, especially those that address the needs of immunization programs in the developing world, are generally viewed as for the public good -- that is, they are less subject to nationalistic and commercial competition and rivalry, and therefore would be more acceptable as a pioneering arena of true international collaboration involving shared funding and shared staffing, with a genuine chance for shared benefits. Regional and international cooperation in public health programs can benefit all countries, regardless of the degree of each individual country's contributions. Intra-regional cooperation in the public health sector would also enhance, not threaten, international or inter-regional cooperation with other national and regional centers.

Such an institute would also be better able to assist public- and private-sector vaccine producers, by providing technical and administrative support through preclinical and clinical evaluation of candidate vaccines, by lending its international prestige and public trust to any new vaccine that may emerge with its assistance, and by aiding the development of technical expertise of national control authorities where such capabilities are not fully developed. Thus, one major role of the institute would be forging strong cooperative partnerships with the commercial sector, in order to lower the tremendous entry barrier that now exists for new vaccines and new vaccine formulations designed for public sector use.

C. Why in the Asia-Pacific Region?

The Asia-Pacific region encompasses the East and Southeast Asian countries and contains more than a third of the world's population. Several countries in the region are major producers of vaccines. Significant producers include Japan, China, Indonesia, South Korea, Australia, North Korea, Thailand and Vietnam. China is in fact the world's largest producer of vaccines, with its total production of all vaccines exceeding 800 million doses a year.

The Asia-Pacific region is currently undergoing unprecedented economic and technological development. These changes have brought with them a surge of national and regional self-confidence and optimism, and a new capacity to contribute regional resources to solve important global issues of the day. In addition, increased knowledge and understanding of the issues in vaccine production and quality control has resulted in heightened interest and commitment on the part of the governments in the region to upgrade vaccine production and quality control and to invest in vaccine-related research and development. The scientific, economic and political climate in the region is therefore conducive to supporting and sustaining a new international center of vaccine-related sciences that will promote cooperative regional and global efforts in research and development, technical assistance and cooperation, and education.

Soon after I submitted the position paper, Frank called me to say that DGIP/UNDP was keenly interested in the proposal, and invited me to meet with his colleagues so that a detailed project plan could be discussed. On June 8, 1992, I went to the UNDP office to meet with Frank, and Timothy Rothermel, Director of DGIP. Frank, a Dutch citizen from Utrecht, had a long career with UNDP in capacity building programs in Africa, Middle East and other areas of the world. Tim Rothermel was an American lawyer, who had a strong personal commitment to helping developing countries build institutional capacities. I also met with Michael Sacks, an American physician and a long-time veteran of several UN agencies, who was a senior advisor to DGIP at that time. The three men were not the typical institutional bureaucrats. Rather, they were practical and informal, action-oriented people with a broad view of UN's role in empowering the developing countries.

Frank said my proposal for IVI represented an innovative way to mobilize the capabilities of the developing countries and bring new players and new funding into the vaccine field. Frank told me that they had already agreed among themselves to support my proposal essentially as I had submitted it, without modifications or revisions, as a formal UNDP project. They believed that the IVI Project will fit perfectly into UNDP's mission of building institutional capacity in vaccines in developing countries, and that it would serve as a major contribution to CVI by UNDP. Then I was asked to consider joining UNDP for a year and lead a feasibility study to test the viability of the IVI idea. I was quite startled because the invitation was quite unexpected.

On July 1, UNDP sent me a formal letter, inviting me to serve as an expert consultant to UNDP starting on September 1. To my surprise, UNDP simply assumed that Samsung Group could just "lend" me to UN for a year to carry out the study. It was of course not possible for me to take a one-year leave from ETI/Cheil for the UN assignment; I would have to resign from the company if I accepted the invitation. Accepting the job at UN would represent a radical change in my life. A one-year assignment for UN without a guaranteed job tenure would be a risky career move for me. In addtion, the daily commute to the UN office on Manhattan's Eastside from my home in Northern New Jersey could be rather stressful.

On the other hand, IVI was my idea, and I would be betraying my own ideals if I refused the opportunity to realize the idea without a convincing reason to myself. In addition, becoming a member of the United Nations would be an honorable and worthy career challenge for me. After a month of hesitation, I decided to take on the risky adventure, and informed UNDP that I will accept their invitation, but starting in October. I was assured that my assignment would be for one year initially, but could be extended as necessary.

On October 1, 1992, I reported to work at my new office at UNDP on East 45th Street in Manhattan, with the title of Senior Health Advisor. I signed the "Oath of Office" to the United Nations, pledging that I will hold the high ideals of the UN above the civic duties to my country of citizenship. I was issued the UN Staff Pass that allowed me free entry to the UN buildings, including the General Assembly Hall and the Security Council Chamber, and received the blue UN passport. I became an international civil servant of the United Nations. My job as a UN officer would eventually last 7 full years, to September 1999.

3

Feasibility Study for IVI Initiated by UNDP

My mission for the feasibility study at UNDP was to determine whether the East Asian/Western Pacific countries would support the basic premises underlying the proposal for IVI, and whether any of the countries would be willing to make the financial and human investments to host and support it.

UNDP sent out formal request to each UN member country in the region, requesting their assistance for the feasibility study by scheduling meetings for me with appropriate officials and scientific leaders. The mission and underlying philosophy of the proposed vaccine institute, as given in the Terms of Reference for my proposed meetings in each country, was summarized as follows:

International Vaccine Institute, a UNDP Project under the

Umbrella of CVI

1. IVI's mission is to promote the health of people in developing countries by the development, introduction and use of new and improved vaccines through a dynamic interaction among science, public health and businesses.

2. IVI represents a new paradigm of development and capacity building for developing countries. It is a mechanism to mobilize their latent resources in order to strengthen human and institutional capacity in development, regulation and use of essential vaccines for the benefit of all.

3. IVI represents a radical change from the externally-driven development assistance mode to a self-driven initiative of the developing countries working in global partnership.

Because this was an initiative by UNDP, many high-level doors were readily opened for me, so I was able to quickly schedule meetings with senior officials and leading vaccine scientists in Bangladesh, China, Hong Kong, Indonesia, Japan, Malaysia, Myanmar, Philippines, Singapore, South Korea, Taiwan, Thailand and Vietnam. I also visited institutions involved in vaccine research and production in Cuba and India to survey and collect information on international cooperation programs in vaccine research that may be helpful for organizing a vaccine institute in Asia. In many of these countries, I was able to meet with scientific leaders and public health officials whom I had come to know personally through my previous work with hepatitis B vaccine programs.

By the middle of 1993, I completed my first round of consultations by visiting all of the Asian countries on my list and submitted a detailed report on each of the countries. However, my visit to Taiwan caused a bit of a stir. Due to my political naivete, I went to Taiwan and met with the Health Minister to present the IVI project and to invite Taiwan's participation and support. From my previous contacts with several senior scientists at the Academia Sinica of Taiwan and National Taiwan University, I felt that Taiwan was one of the best-qualified countries in Asia to host IVI. After my visit to Taiwan, China officially objected, saying that Taiwan was part of China, not an independent member state of the UN, and thus could not participate in a UN-initiative. The visit became my first and last visit to Taiwan in connection with the IVI project.

Frank Hartvelt and I also visited vaccine-related institutions in Egypt and Brazil, to explain the IVI initiative and to invite them to join in a future coalition of vaccine centers which we envisioned for IVI as part of its global role. Egypt was said to have been a significant vaccine producer at their Pasteur Institute, even though its role as a regional vaccine producer was now minimal. Brazil was the largest and most active vaccine producing country in South America. We met with the leaders of the Oswaldo Cruz Institute in Rio de Janeiro and the Butantan Institute in Sao Paolo, and also with government officials in Brasilia. Separately, I also visited officials and vaccine institutes in Mexico, Cuba, Chile, and India.

Through these consultations, I was able to confirm rather quickly that there was a surprisingly broad support for the IVI idea in Asian countries. At least 6 of these countries were interested in hosting the institute, because it was seen as an opportunity to obtain international assistance to develop their domestic vaccine-related capabilities and to boost their research base in health sciences in general. I made it clear that even though UNDP may provide a significant part of the cost during the initial phase of IVI's operation, the host country should be prepared to pay for the construction of the institute headquarters building and laboratories, as well as for a major part of the operating costs thereafter.

In August 1993, based on my final report of the feasibility

study, UNDP initiated the steps that were required to formally establish IVI as an international organization. In November, UNDP's proposal for IVI was approved as part of the CVI Strategic Plan at the CVI Consultative Group meeting in Kyoto.

At this critical time, I had the good fortune of Gurinder Shahi joining my team. He was a native of Singapore, from a devout Sikh family. After graduating from the medical faculty of National University of Singapore and receiving a master's degree from Harvard School of Public Health, he was working as a staff intern at the Rockefeller Foundation in New York. Gurinder said he wanted to join the IVI project team because IVI represented an innovative example of empowering the third world. He was a superb organizer and strategic planner. With Shahi on the staff, my work to organize the new institute gained speed.

4 / Institutional Politics and Commercial Interests Intervene

Before I embarked on the feasibility study for IVI, I had expected that the four other sponsoring partners of the CVI Coalition (WHO, UNICEF, the World Bank and the Rockefeller Foundation) would enthusiastically welcome the UNDP initiative, since it would be a novel mechanism to bring the potentially large human and institutional resources of the developing countries to the CVI movement. This was what CVI had aimed to achieve, I believed. But I would soon discover that I was too naïve about the institutional politics and inter-organizational rivalries that operated even under the lofty goals of CVI.

Soon after UNDP announced in October of 1992 that it was undertaking the feasibility study for IVI, Frank Hartvelt and I faced a strong opposition from WHO. They objected to the IVI idea because all things related to vaccine development, production, licensure and distribution should be the exclusive purview of WHO. My original version of the IVI Proposal to UNDP had indeed contained the suggestion that IVI could serve as the regional mechanism to help the member countries to collectively test and certify the vaccines developed by them. This of course reflected my own personal experiences with the hepatitis B vaccine in several Asian countries. WHO's position was that vaccine registration should be the sovereign function of individual states, with guidance from WHO. UNDP accepted that position, and revised the IVI Proposal accordingly by deleting regulatory issues from the list of IVI's functional areas.

In the spring of 1993, I went to Manila to meet with Prof. Ernesto Domingo, the Chancellor of the University of Philippines and a strong supporter of the IVI idea, in order to invite the Philippines to join in the IVI program. On that occasion, UNDP sent a request to the Western Pacific Regional Office of WHO (WPRO), which is located in Manila, to schedule a meeting for me, but WPRO did not. When I went to the WPRO office, I was told that the director was not in town that day. The Regional Director at that time, S.T. Hahn, and his chief deputy, J.W. Lee (both of whom were, incidentally, South

Korean citizens) were well aware that South Korea was a strong supporter of the UNDP initiative and a potential host of IVI. A few years later, when J.W. Lee became the Director of Vaccines and Immunization of WHO in Geneva, he admitted to me that he was personally responsible for blocking the meeting with the director because WPRO was strongly opposed to the IVI project.

UNDP believed that the IVI project was essentially a capacity building program that would complement, not compete against, the vaccine-related activities of WHO. We therefore took steps to assure WHO that IVI will not "invade" their territory, by announcing that IVI will not be a producer of vaccines, and will not be a regulatory agency. Further, UNDP promised to assign two seats on the IVI's Board of Trustees to WHO (one for WPRO and another for WHO Geneva), even though UNDP will assign one seat for itself. Eventually, in 1997, WHO was one of the first signatories of the IVI Establishment Agreement.

Japan was an important partner of the CVI coalition that did not openly support the IVI initiative, even though several prominent leaders of vaccine research and development in Japan warmly endorsed the project. I first met with Prof. Konosuke Fukai, Professor Emeritus and Chairman, and

Prof. Michiaki Takahashi, Professor Emeritus and Deputy Chairman, of the Research Foundation for Microbial Diseases (Biken) of Osaka University. They were Japan's most renowned and respected leaders of vaccine research and development. Dr. Fukai was an advocate for strengthening local vaccine manufacturing capabilities in Asian countries, and had led a program to help vaccine research and production in Vietnam. Dr. Fukai told me that an international vaccine institute to help other developing countries in Asia was an excellent idea, but such an institute should be located in a country such Korea or Vietnam, not in Japan, because Japan already had well-developed vaccine industry. I also visited the Kitasato Institute and the Japan National Institute of Health. The senior scientists whom I met at both institutions warmly endorsed the IVI initiative.

However, Japanese government institutions did not formally endorse the UNDP project. The Director General of WHO in Geneva at that time was Hiroshi Nakajima, and WHO's widely-recognized opposition to the UNDP initiative probably prevented Japan's support.

Another member of the CVI Council that did not fully support the IVI initiative was the Rockefeller Foundation. Scott Halstead, who represented the Foundation at CVI, said that the global community should not invest "in bricks and mortar" for a new institution to be located in a developing country, but should use the money for more practical programs (which would naturally mean programs in the developed countries). The reactions from UNICEF and the World Bank could be best described as neutral.

Soon enough, I would come to see that the conflicting positions regarding UNDP's IVI initiative shown by some of the coalition members of CVI were most likely the reflection of political jockeying and institutional rivalry that became evident with other issues of the CVI coalition. William Muraskin, the social historian, provides an interesting historical analysis of the prevailing situation at this time, in his penetrating book, *"The Politics of International Health: The Children's Vaccine Initiative and the Struggle to Develop Vaccines for the Third World"* (State University of New York Press, 1998). In the book, Muraskin devotes a separate chapter to describe the birth of IVI within the CVI context, *"The International Vaccine Institute: A CVI Spin-off Pioneers a More Aggressive Path"*.

The CVI movement represented a broad international coalition, and included representatives of large multinational vaccine companies as well. Initially, the multinational vaccine companies were reluctant to endorse the IVI project, because of their openly-expressed concerns about intellectual property rights. They were probably also worried about potential competitions from new low-cost players in the vaccines business that IVI could promote.

At this time, we became aware that a mid-level executive of a large American vaccine company sent a covert letter to the U.S. Department of Commerce, stating that IVI would help its partners "steal" American vaccine technology and therefore the U.S. should not support the IVI project. Most probably because of the covert letter, which was written even before any of the research programs of IVI were fully announced, the U.S. did not sign the IVI Agreement. Today, in 2023, 39 member states of the UN and WHO have signed on the IVI Agreement. Successive directors of the Institute, except for a brief interlude, were all American citizens, as were many of the senior-level staff. The U.S. is still not a signatory to the Agreement.

In response to the reservations and misconceptions expressed by some of the CVI partners, UNDP announced publicly that IVI will not produce vaccines for commercial sale, and that IVI will promote and protect intellectual property rights. UNDP also invited Maurice Hilleman of Merck Sharpe and Dohme to serve on the Interim Board of Trustees of IVI. Dr. Hilleman is the pre-eminent vaccine scientist of the 20th century who personally led the development of many vaccines. Dr. Hilleman accepted the invitation, and served on the Interim Board when it was officially convened in 1995.

In view of the enthusiastic support from the Asian countries, UNDP concluded that the IVI project should move ahead. Frank Hartvelt and I were convinced that our position was justified. We believed that those who expressed reservations failed to understand that the developing countries in Asia would bring in "new" resources into the CVI movement, rather than depending on the funds from the richer countries. Personally, I also thought that there was perhaps a "historical and cultural bias" against the idea that developing countries themselves could take the leading role in developing and producing new vaccines that their populations desperately needed.

My conviction that our position was justified was based on my finding that most leading scientists and institutions that were not bound by bureaucratic self-interests enthusiastically supported the IVI idea. Also helpful for UNDP was the show of support from senior scientists affiliated with the U.S. government institutions, most notably John La Montagne, Deputy Director of the National Institute for Allergy and Infectious Diseases, and Philip Russell, the former Director of the Walter Reed Army Institute of Research who was at that time serving as the Special Advisor for CVI.

Leading scientists in many countries in Asia, Africa and Latin America, as well as institutions that I have consulted in UK, Sweden, Norway, Finland, the Netherlands, France and Israel, generally welcomed the IVI initiative as well. As the specific programs of IVI became more widely known, individuals and organizations that were once reluctant to support IVI, such as the Rockefeller Foundation and PATH, gradually became supporters and collaborators. Eventually, major U.S. and European vaccine companies also came to view IVI as a helpful collaborator rather than as a competitor.

5

Host Country Selection Becomes an International Competition

In September of 1993, UNDP formed a planning group for IVI, and took steps to select the host country. We knew that several Asian countries were strongly interested in hosting the Institute, and therefore anticipated a strong competition among them. It was thus essential that the selection process was impartial and transparent.

In October, UNDP named a Site Selection Committee to oversee the selection process. The Committee was composed of prominent international experts who represented major geographical areas and scientific disciplines, as follows:

Membership of the Site Selection Committee (1993-1994)

- Ms. Margaret Catley-Carlson, Chairperson (Canada) (President of Population Council, New York; Former Minister of Health of Canada)
- Dr. Demissie Habte (Ethiopia) (Director General, International Center for Diarrheal Diseases Research, Dhaka, Bangladesh)
- Dr. John La Montagne (USA) (Deputy Director, National Institute of Allergy and Infectious Diseases, Bethesda)
- Dr. Geoffrey Schild (United Kingdom) (Director, National Institute for Biological Standards and Control, London)
- Dr. Gurusaran Prasad Talwar (India) (Director, National Institute for Immunology, New Delhi)
- Dr. Philip Russell, Ex Officio, Special Advisor to CVI (USA) (Johns Hopkins University, Baltimore)

Gurinder Shahi and I compiled a detailed 100-page document, called "*Document File for Members of the Site Selection Committee,*" dated October 8, 1993. The book contained the Feasibility Study Report that I had submitted to UNDP, with my meeting reports on each country and the correspondences with country representatives confirming their desire to host

IVI. It also contained the proposed schedule of dates for the initial reviews, site visits to short-listed countries, and the final selection.

With the approval of the Site Selection Committee, UNDP sent invitations for application to host IVI to 7 countries (China, Indonesia, Malaysia, the Philippines, Singapore, South Korea and Thailand) and the government of Hong Kong on November 30, 1993. The invitation was a 41-page document, titled *"International Vaccine Institute: Invitation for Submission of Proposal from Prospective Host Countries."*

The *Invitation* listed in detail the key information that the Site Selection Committee would use to select the winner, including the following points:

1. The application must have the official endorsement of the national government, and must designate a specific host city for IVI.

2. The host country should commit to providing a purpose-built headquarters building that includes state-of-the-art research laboratories, training facilities and a pilot plant. The host country will also provide at least 30% of the annual operating cost of IVI.

3. The host country will ensure that IVI will have the privileges and immunities appropriate for an independent international organization.

The deadline for submission of the application was March 15, 1994. The Site Selection Committee will first select 3 countries based on the review of the applications, and then visit the short-listed countries for on-site evaluations in April. The final selection of the host city was scheduled for June 1994.

In order to encourage the countries to respond to UNDP's invitation and to clarify any issues regarding the selection process, I traveled once again to Beijing, Seoul, Hong Kong, Jakarta, Manila, Kuala Lumpur, Bangkok and Singapore, and met with the key leaders in charge of their bid for IVI. I was aware that not every country had firmly decided to submit an application yet, and that internal debates were still going on in some of them.

In my final meetings with the leaders of Asian countries, I framed UNDP's proposal for IVI as a simple, idealistic message that the time has come for Asian countries to take a more proactive role in global public health affairs, and that the IVI initiative by UNDP provided a unique opportunity to do so, as follows:

International Vaccine Institute is a Regional Development Initiative of the Asia-Pacific Countries

1. The Need

The Asia-Pacific Region contains countries in critical need of international assistance to achieve greater vaccine self-sufficiency.

2. The opportunity

The Region as a whole possesses sufficient resources in

Scientific, technical and industrial manpower

Social and administrative infrastructure

Economic resources

Dynamic leadership

that are matched by political commitment and a sense of "historic moment" to contribute to global issues.

3. The Challenge

How to channel the Region's latent resources towards the goals of the global CVI movement, and at the same time help accelerate the economic and social development.

Below are excerpts of the relevant portions of the situation reports for each country that I wrote at the end of my Feasibility Study for IVI. The country reports for Vietnam and Malaysia are not included here, because these two countries informed UNDP in advance that they decided not to submit an application.

China

My first visit to Beijing was in late 1992, to meet with Prof. Zhu Zhiming, to brief him on the IVI initiative and to invite China's support and participation. Prof. Zhu was the highly respected doyen of Chinese health scientists, and the senior mentor to the new generation of leaders, including Prof. Chen Chunming, the Founding President, and Prof. Wang Ke-an, the Deputy President, of the Chinese Academy of Preventive Medicine. I knew from my British colleagues that Dr. Zhu was considered a rising star in virology when he was a young researcher at Cambridge University. He returned to China in the early 1950's, soon after Mao Zedong unified China, saying that he wanted to contribute to the development of science in his home country. He later founded the National Institute of Virology located in the Tiantan District of Beijing, where he still had an office and where I first met with him.

Prof. Zhu convened a large briefing meeting for me on October 28, which was attended by a dozen senior members of the National Institute of Virology, Chinese Academy of Preventive Medicine, China National Committee for Biotechnology Development, National Vaccine and Serum Institute of Beijing, and National Institute for Control of Pharmaceutical and Biological Products. Resident representatives of WHO, UNICEF and UNDP based in Beijing also attended the briefing meeting.

At the end of several private conversations with me following the formal meeting, Prof. Zhu said that China was not yet ready to undertake a major international project such as IVI, because he believed that Beijing did not have the physical infrastructure nor the scientific manpower needed for it. It was true that, at that time, the main 4-lane roadway from the Beijing Airport to the city center was not an expressway and meandering cows sometimes blocked the traffic. The main city boulevards around the Sheraton Hotel in Chaoyang District, the up-scale area where many of the foreign delegations were located, were densely packed with commuters on bicycles, not by cars.

I pointed out that China, even in 1992, was probably the largest vaccine producer in the world in terms of the number of vaccine doses produced. There were well-established vaccine production centers in major cities such as Beijing, Shanghai, Chengdu, Changchun, Kunming and Lanzhou. China had a longer production history and deeper scientific experience in vaccine development than any other country in East Asia, except for Japan. With a huge population and corresponding needs for children's vaccines, Beijing would be a good location for IVI.

When, in late 1993, I went to Beijing for the third time and met with Prof. Zhu again, he told me that China has finally decided to compete for IVI. He was persuaded that China can mobilize the political will and financial resources to support IVI and, through the UNDP initiative, contribute to other developing countries. He took me to a vast empty field in the outskirts of Beijing, about half an hour from the Tiananmen Square, and told me that the area was going to be developed as a cutting-edge science park, and China will propose to locate IVI there. Indeed, Beijing was on the cusp of becoming a modern metropolis. Physical signs of the transformation were visible everywhere, with new expressways and high-rise buildings going up in every corner of the enormous city.

True to the long Chinese academic tradition, Prof. Zhu was deeply respected by the younger generation of scientists, and he had their enthusiastic following and support. He asked Prof. Wang Ke-an to organize Beijing's bid for IVI. Prof. Wang quickly assembled a team of bright, idealistic young scientists, many of whom drawn from the Institute of Virology and the Chinese Academy of Preventive Medicine, to prepare China's application. I knew that Beijing's application for IVI would be one of the most competitive.

South Korea

I first went to Prof. Wan-kyoo Cho in Seoul in December 1992 to brief him on the UNDP initiative for IVI, and requested that Korea consider hosting the institute. Prof. Cho is a respected elder statesman of the Korean scientific community who would be the natural leader for a Korean bid for IVI. He had served as the President of Seoul National University and as Minister of Education, in addition to many other positions related to promoting Korean science and education. Under Prof. Cho's leadership, a strong consensus quickly emerged to actively seek to host IVI in Korea.

When I visited Korea again in April 1993, a group of very senior leaders from the academic, medical, government and industry circles was convened for a meeting with me and

Jacob Guijt, the resident representative of UNDP in Korea. Prof. E-Hyock Kwon, former President of SNU and Minister of Health, and Prof. Howang Lee, a world-renowned virologist and discoverer of the Hantavirus, were among them. The group emphasized to us that the Korean scientific community strongly supported bringing the institute to Korea. Also present was Young-Sup Huh, Chairman of the Bioindustry Association of Korea and CEO of Green Cross Corp, a major vaccine company. He studied in Aachen, Germany, and was a champion of increased international engagement by Korea. Like many others of their generation, these leaders believed that Korea had a moral obligation to help the poorer nations of the world, just as Korea was helped by other countries when Korea was a poor nation devastated by war. In addition, IVI could be a boost for Korean health sciences in general, and especially for the biotechnology industry.

Soon after my second visit to Seoul, the Korea IVI Organizing Committee was created to respond to the UNDP initiative. The Committee was a broad-based group covering the academia, government and industry, and Prof. Wan-kyoo Cho was elected as the Chairman of the Committee. Its membership also included young faculty members of Seoul National University, most of whom trained in American universities and eager to internationalize Korean science by bringing IVI to Korea. It was obvious that members of the Committee shared the idealistic vision that Korea, which for long was on the receiving end of international assistance, especially after the Korean War, could now be on the giving end by contributing to vaccine development for the poorer countries of the world.

Prof. Sang-dai Park, Dean of Research Affairs of SNU at that time and a former student and close associate of Prof. Cho, took the leadership of a working group of young professors charged with preparing for the Korean application. The group had enthusiastic support from Prof. Chong-un Kim, the SNU President.

In January 1994, a delegation of the Korea IVI Organizing Committee, composed of Prof. Cho, Prof. Park and Prof. Rho-hyun Seong of SNU, visited UNDP office in New York and again confirmed that Korea will submit an application to host IVI.

Compared to Hong Kong and Singapore, and perhaps even to Bangkok, Seoul in 1994 could not have been called a polished cosmopolitan city that it is today. The housing stock, public transportation systems and general living environment in Seoul were not friendly to foreign residents who did not speak Korean. At that time, Seoul was known for horrendous traffic congestions and severe air pollution, especially in winter. But

I could see that the Korea Organizing Committee was the best organized group, and that Seoul probably had the strongest support base among the competing Asian cities.

Hong Kong

Hong Kong's effort for IVI was led mostly by Prof. Stephen Chung, a molecular biologist who was trained at Berkeley and MIT and now directed the Institute for Molecular Biology of Hong Kong University. He was also an old colleague of mine because he was a senior scientist at Eugene Tech International, the biotech company in New Jersey that I founded.

In January 1993, I visited Hong Kong to meet with Prof. Chung and other Hong Kong scientists who supported Hong Kong's bid to host IVI. Stephen said he was hoping to use the Hong Kong Institute of Biotechnology as the anchor for IVI. The Institute was housed in a newly built but under-used research center on the waterfront in the New Territories, near the campus of the Chinese University of Hong Kong. Stephen said Hong Kong certainly had the financial resources to support a new international scientific institution, and Hong Kong's academic community had a strong desire to develop the city as a center for scientific research. IVI would be an ideal fit. Another very important consideration for them was that locating an international organization created by a UN agency would strengthen Hong Kong's future status as a free, independent city even after the control of the city returned to China, which was scheduled for 1997.

However, I sensed that Hong Kong's scientific community was having difficulty securing the political and financial commitments for the IVI project from the Hong Kong government. In an effort to persuade officials of the British-run government, Stephen organized a meeting for me with Prof. Charles Kao, President of Chinese University of Hong Kong, Prof. Tim Biscoe, a British citizen and President of Hong Kong University, and Dr. Lee Shiu-Hung, Director of Health Administration and Planning of the Department of Health of Hong Kong Government. I also met with several other senior members of the city who, according to Stephen, had great influence in official policy making.

As a follow-up to the first round of meetings in Hong Kong, Prof. Kao of Chinese University of Hong Kong made a special visit to UNDP in New York on July 15, 1993, and met with Tim Rothermel, Frank Hartvelt and me. He wanted to convey the strong wish of Hong Kong's scientific community to host IVI. But the British-controlled government officials seemed to be preoccupied with the impending transfer of Hong Kong's control back to China, only four years away. They did not come forward with the political commitment that was necessary for Hong Kong's bid for IVI.

For many reasons, I believed that Hong Kong would have been an excellent host city for IVI. It was a modern, English-speaking cosmopolitan city that had a long history of being open to the world. The city was eager to become a center of scientific research as well, as shown by its support for the newly established Hong Kong University of Science and Technology that matched any modern American university campus. Hong Kong would be an attractive place to work and live for IVI's international staff and their families. In my view, Hong Kong could be in many ways the best city to host IVI.

To my surprise, however, Hong Kong did not submit an application for IVI by the deadline of March 15, 1994.

Thailand

In Thailand, the effort to host IVI in Bangkok was led jointly by Dr. Prayura Kunasol, the Director-General of Department of Communicable Diseases of Ministry of Public Health, and Prof. Natth Bhamarapravati of Mahidol University who was President Emeritus of the University and Chairman of the National Vaccine Policy Committee of Thailand. He was a well-known scientist engaged in Dengue vaccine research, and had established the Center for Vaccine Development of Mahidol University at Salaya, a suburban town near Bangkok. In fact, IVI project almost seemed to be a personal crusade by Prof. Natth. He invited me to visit the Center that was housed in a laboratory building with experimental animal facility. It was located in a rural setting, with more land around it. Prof. Natth said he hoped to turn the vaccine center facilities as the core of the new IVI if Thailand became the host city. Dr. Natth's plan had the support of Dr. Kunasol and other public health officials when I met with them for the second time on May 3-4, 1993, in Bangkok.

Many international organizations were based in Bangkok, and the cosmopolitan city was a favorite destination for international travelers. Bangkok probably lacked the strong scientific infrastructure and broad-based political and academic support for hosting IVI that the other potential competitor cities had, but Bangkok was certainly a strong candidate.

Singapore

My first briefing meetings in Singapore were held on January 19-20, 1993, at National University of Singapore. It was a large meeting chaired by Prof. Christopher Tan, and attended by more than a dozen senior members of the government and academia. Separately, I also met with a small group of government officials and briefed them on the UNDP initiative.

The leading figures who were preparing Singapore's response to UNDP's call were Prof. Tan and Teoh Yong Sea. Prof. Tan, a molecular biologist trained in Canada, was the key figure who founded the Institute of Molecular and Cell Biology (IMCB) on the campus of the National University of Singapore in 1985. He shepherded the new Institute to become one of the best molecular biology laboratories in East Asia in a very short time. I was impressed when I first visited IMCB, which was housed in a well-designed modern laboratory building on the leafy grounds of the National University. T.Y. Sea was Director of the National Biotechnology Program at Economic Development Board, the main government office that managed the industrial development programs of Singapore.

Dr. Tan said they wished to bring the UNDP-sponsored institute to Singapore because the government was interested in developing biotechnology as a key industry and vaccine was an area of interest. He said he was proud of the fact that Prime Minister Lee Kwan Yew was a strong supporter of science and technology development in Singapore. The Prime Minister was personally involved in establishing and supporting the new IMCB. He would be interested in the IVI project as a means of promoting science and technology development. Chris Tan appointed Hugh Purser, a senior staff at IMCB, as the coordinator for Singapore's application for IVI.

On April 22, 1993, a Singapore delegation headed by T.Y. Sea visited UNDP in New York for a follow-up meeting with me, Frank Hartvelt and Tim Rothermel. They presented the reasons why Singapore was interested in hosting IVI, and why Singapore would be an ideal host country for it. One important point they wanted to empathize was that Singapore had the capacity to provide appropriate resources "in cash and kind to support the Institute" if selected.

There were many reasons that would make Singapore an attractive place for IVI. English was the official language, and the city state had an easy international access. As a small country with an open society and neutral politics, Singapore was a place where people from all countries would feel welcome. The modern and efficient city had a highly developed urban infrastructure and healthcare facilities for the staff and their families. Research and development of new vaccines would accord well with Singapore's ambition to promote the biotech industry as a national policy. Singapore would be a very competitive candidate for IVI.

Indonesia

Indonesia expressed its support for IVI and a strong desire to bid for it from the very beginning of the UNDP feasibility study. At my first briefing meeting in Jakarta on January 22, 1993, several senior Indonesian leaders were present, including Prof. A. Loedin, Assistant Minister of State for Research and Technology, Dr. Sangkot Marzuki, Director of the Eijkman Institute for Molecular Biology, and Drs. Darodjatun, President Director of Perum BioFarma (Pasteur Institute), a major vaccine company in Bandung. They were joined by several other heads of research labs and government institutions.

Indonesian bid for IVI seemed to be led by two different groups, one in Jakarta and another in Bandung. Prof. Marzuki invited me to tour the newly established Eijkman Institute in Jakarta. He said he hoped to make the Institute the anchor for IVI. Drs. Wim Kalona, President Director of Darya-Varia

Laboratoria, a producer of veterinary vaccines and other biologicals in Jakarta, strongly endorsed Prof. Marzuki.

Drs. Darodjatun was interested in inviting IVI to Bandung. He was an energetic and charismatic leader who made BioFarma the largest and most important player in vaccine production in Southeast Asia. Bandung was the center of Indonesian high-tech industry, where the government was actively promoting an ambitious national program to develop its own airplane industry. Drs. Darodjatun believed that BioFarma would serve as a good partner for IVI. Since Bandung is located high on a mountainous plateau in West Java, it enjoys a moderate climate compared to the tropical city Jakarta, which was the reason the city of Bandung was developed during the Dutch period and still favored by many foreign residents. Bandung would be a good city to host IVI.

The Philippines

Prof. Ernesto Domingo, Chancellor and former Dean of Medical School of the University of Philippines, was an eager supporter of IVI and the leader of his country's effort to bring the institute to Manila. He was a soft-spoken scholar who desired to promote vaccine research and development in the

Philippines. He had become a close personal friend of mine because he and I worked together to bring the hepatitis B vaccine to the Philippines.

On November 3, 1992, Prof. Domingo organized a meeting for me with himself, Dr. Jaime Tan, Senior Undersecretary of the Ministry of Health, and Felipe Miranda, Professor of Political Science and Policy Advisor to the President of the Philippines. Officials from the WHO-WPRO Manila office and from the International Rice Research Institute of Los Banos also attended the meeting. Everyone strongly endorsed the IVI initiative, and hoped to bring it to Manila.

Since the days when Manila was run by the Americans, the city was home to many international organizations, including the Western Pacific Regional Office of WHO. IVI could easily fit into Manila's international environment. But it seemed that Ernesto was not able to mobilize a broad-based national support for the IVI project since it required the government to commit a large sum of money and human resources in advance. Manila did not have significant local organizations involved in vaccine research or vaccine production, and in the end failed to organize an effective working group to produce a strong application for IVI.

6

Seoul Wins
the Competition for IVI

Six countries submitted applications to host IVI by the deadline of March 15, 1994. They were China, Indonesia, Korea, the Philippines, Singapore and Thailand. The Site Selection Committee selected Beijing, Bangkok and Seoul as the three finalists for further on-site reviews. Surprisingly, Singapore was not on this list. Singapore said they wanted to host IVI, but with a somewhat reduced scope of activity, mostly focused on vaccine development research, with the condition that any intellectual property generated at IVI should belong to Singapore. These conditions were not acceptable to UNDP.

In May, the Site Selection Committee members visited Bangkok, Seoul and Beijing, in that order, for a 2-day on-site review in each city. The selection process had become an international competition by the 3 cities. UNDP wanted to ensure that the process remained impartial and transparent, and was seen as such by everyone involved. So I did not join the on-site review team because my participation could have been construed as unfairly lending support for Seoul. Gurinder Shahi accompanied the Committee members as the UNDP staff.

On June 27, 1994, the Site Selection Committee met in a closed-door meeting at UNDP in New York to make the final selection. We learned afterwards that the Committee used a check list of 35 points and graded each city against it. The list covered a long catalogue of points, ranging from the strength of the political and financial commitment of the national government to support IVI, the availability of local scientific manpower and the breath of in-country vaccine-related activities, to the quality-of-life considerations for the foreign staff and their families in the host city, such as the availability of housing, international schools and medical care. At the end of the day-long session, Margaret Catley-Carlson, the Chairperson of the Committee, informed UNDP that Seoul was the final winner.

In a letter dated June 28, Timothy Rothermel, Director of the Division for Global and Interregional Programmes of

UNDP, informed Ambassador Chong-Ha Yoo, Permanent Representative of the Republic of Korea at the United Nations, that Seoul was selected as the host city for IVI. The letter also contained the following two statements:

"...Based on these considerations, the Committee recommended to UNDP that the location of the IVI be in Seoul, Republic of Korea, with a major training facility located in China...

...In view of the very strong support and commitments of the countries of the East Asia and the Western Pacific to the objectives of the institute, and in view of the broad range of capabilities that exist in the region, the Site Selection Committee strongly recommended also that a network of cooperating facilities, joined together in common purpose, be established. UNDP hopes to implement these recommendations as far as possible..."

The designation of the Republic of Korea as the host country of the International Vaccine Institute and the reaffirmation of the Korean government's support for it were officially reconfirmed in September by an exchange of letters signed by James Speth, the Administrator of UNDP, and

Ambassador Yoo, the Korean Permanent Representative to United Nations. Seoul was now officially the host city for IVI.

UNDP prepared for the opening of the IVI office in Seoul. There were additional formal steps to complete, including the drafting of the Agreement on the Establishment of the International Vaccine Institute, which was an international treaty for establishing IVI as an autonomous international organization under the Vienna Convention, and the Constitution of IVI, its governing charter. UNDP and the Republic of Korea agreed on the text of the Establishment Agreement, with a Preamble and eleven Articles, and opened it for signature in October 1996.

The central mission of the Institute was re-stated as follows:

To accelerate the introduction of vaccines into developing country public health programs by undertaking research and providing research-based technical assistance that effectively address issues of vaccine development, disease burden, safety and efficacy, delivery feasibility and effectiveness, and sustained supply.

An Interim Board of Trustees of IVI was constituted by UNDP. The Board would serve as the advisory body to UNDP while the institute remained a UNDP organization, but would later become the permanent Governing Board when IVI gained the formal status as an international organization and separated from the United Nations.

UNDP invited 3 eminent international leaders to serve as the Board Nominations Committee. They were Dr. Nyle Brady, Chairman (USA) (Professor at Columbia University, and former director of the International Rice Research Institute in the Philippines), Dr. S. Ramachandran (India) (Former Minister of Biotechnology of India), and Prof. Hans Wigzell (Sweden) (President of Karolinska Institute, Stockholm).

UNDP invited UN member states to nominate candidates for the Board, and received about 90 nominations from 35 states. From the list of candidates, the Committee elected the following *at-large* members for recommendation to UNDP. With such a distinguished group of scientists and public health leaders as the inaugural members of the Board, IVI instantly gained a degree of prestige and recognition unusual for a nascent institution.

Members of the Inaugural Board (1997)

- Prof. Barry Bloom, Chair of the Board (Chairman, Department of Microbiology and Immunology, Albert Einstein College of Medicine, New York, USA)
- Prof. Gustav Nossal, Vice Chair of the Board (Director, Walter and Eliza Hall Institute, Australia)
- Dr. Ruth Arnon (Vice President for Scientific Relations, Weizmann Institute of Science, Israel)
- Prof. Chen Chunming (President, Chinese Academy of Preventive Medicine, China)
- Prof. Wan-kyoo Cho (President, Korean Academy of Science and Technology, Korea)
- Drs. Darodjatun (President Director, BioFarma, Indonesia)
- Prof. Demissie Habte (Director, International Center for Diarrheal Diseases Research, Bangladesh; Ethiopia)
- Dr. Maurice Hilleman (Director, Merck Institute for Therapeutics Research, USA)
- Dr. Adolfo Martinez-Palomo (Director-General, Center for Research and Advanced Studies, Mexico)
- Dr. Lars Pallesen (Executive Director, State Serum Institute, Denmark)
- Prof. V. Ramalingaswami (Professor Emeritus, All India

Institute for Medical Sciences, India)

- Dr. Geoffrey Schild (Director, National Institute for Biological Standards and Control, UK).

The first meeting of the Interim Board was convened at UNDP in New York in April 1995. These members of the Board continued to serve on it when the Interim Board became the Governing Board of Trustees in 1997, except for Dr. Maurice Hilleman, who resigned in 1996 for personal reasons.

In addition to the elected members, the Board also had additional members named by the host country (2), UNDP (1), WHO/CVI Geneva (1), and the two regional offices of WHO (Western Pacific, WPRO, and Southeast Asian, SEARO) (2).

7

IVI Opens
Interim Office in Seoul

UNDP and Korea agreed that IVI will open its office in Seoul at the beginning of the year of 1995. I agreed to continue to direct the IVI project, and Gurinder Shahi was appointed as head of program development. In October 1994, Richard Mahoney visited me in New York, and asked to join the IVI team. It was welcome news to me, and a major boost for the team. Rich was a cofounder and Vice President of PATH (Program for Appropriate Technology in Health), a Seattle-based NGO with a long history of working in developing countries, especially in Southeast Asia and Africa. He would bring his long experience and dedication to public service to IVI. Furthermore, he and I had worked together as close collaborators for the Hepatitis B Task Force. Rich was charged with leading institutional development. The three of us -- Gurinder Shahi, Richard Mahoney and I -- became the first official members of the UNDP's IVI Project Team, moving together to Korea.

The Korean application for IVI had stated that the headquarters building of IVI, containing laboratories and a pilot plant, will be built on a newly established research park adjacent to the main campus of Seoul National University (SNU), located in the foothills of the imposing Gwanak Mountain that looms over the southwestern skyline of Seoul. The initial plan was to complete the construction of the IVI building in 1998, and until then SNU will provide interim office space for IVI on the campus. I understood that President Chong-un Kim of SNU, a professor of English literature and an enthusiastic supporter of the IVI project, was instrumental in making the valuable land in the research park allocated to IVI. SNU also made arrangements for the IVI Team members and their families to stay at the Hoam Faculty Center guest apartments that were usually reserved for visiting faculty.

Gurinder Shahi and Richard Mahoney arrived in Seoul in the first week of January 1995. Because of last-minute meetings in New York, I arrived in Seoul a little later, on January 17. I clearly remember that, as my taxi drove to the Hoam Faculty Center from the Gimpo Airport, the breaking news on the car radio was about a major earthquake that had just hit southern Japan. It was the Great Hanshin Earthquake of 1995 in the Osaka-Kobe area, with a magnitude 6.9.

Under the leadership of Prof. Wan-kyoo Cho and Prof. Sang-dai Park, several young faculty members formed an *ad hoc* support group to help the UNDP team settle down in Seoul. Prof. Jeongbin Yim, Director of the Institute for Molecular Biology and Genetics of SNU, offered space at his institute for IVI to use as temporary office. After several weeks, the IVI team moved to a large suite of offices, occupying an entire floor of a new building that the university had just opened on the campus. SNU made the space available free of charge, and IVI stayed at this space until the newly built IVI headquarters building was opened in 2003.

We moved quickly to set up the organizational structure by recruiting staff for administration. The first senior staff to join IVI in Korea was Sae-joong Kim as the head of administrative operations. He had served as senior executive for international operations of the Hyundai Corporation. One important responsibility for SJ Kim was to interface with the Korean Government. Ms. Sun-Young Min soon followed as Administrative Assistant.

One important early task for me was establishing channels

of communication and working partnerships with relevant Korean government offices. Another was establishing contacts with members of the Korean academia. We also initiated building the regional network of institutions in Asian countries active in vaccine research and development, a key mission of IVI.

Step by step, IVI was becoming a functioning institution with expanding networking groups in Korea and in several Asian countries.

8 / **Conflicts and Difficulties of the First Years**

During the first two years, we faced several difficult issues, some that we had expected and some that we had not. International institutional politics, in particular the relationship with WHO, continued to simmer even after IVI had opened its office in Seoul. UNDP addressed these issues where appropriate. For instance, because of WHO's concerns about IVI potentially providing vaccine regulatory functions to developing country vaccine manufacturers, we eliminated these functions from IVI's proposed list of activities. UNDP also agreed to assign two seats on the Board of Trustees to WHO appointees, while only one seat was assigned to UNDP appointee, so that WHO could make sure that the operations of IVI would follow these guidelines. We believed that only time and records of impartial operations would eventually solve their concerns.

Quite unexpectedly, however, we also had financial and political problems as well, from UNDP and from Korea. When I first joined UNDP to carry out the feasibility study for IVI, I was given to understand that UNDP would be able to contribute a "meaningful" amount of funding support for IVI, at least enough to cover a significant portion of the initial cost of operation. During my meetings with the leaders of Asian countries who were preparing to submit the application to host IVI, I hinted that UNDP would be a significant initial funding partner, even though the details were never discussed.

However, in 1993, the Clinton-government named James Speth as the new Administrator of UNDP. The Administrator is the head of UNDP and has the rank of Deputy Secretary General of the United Nations. Traditionally, this post has been filled by an American nominated by the US government. The new Administrator drastically cut the program budget for the Division for Global and Interregional Programs (DGIP), which was responsible for the IVI project. The budget cut could have been for internal budgetary reasons or due to institutional politics; I did not know. As a result, DGIP found itself suddenly unable to contribute money to the IVI project except for the personnel cost of the UNDP team. My team was put in a very difficult situation. The Korea IVI Committee had expected a significant contribution from UNDP, especially during the early period when other funding mechanisms were not yet in place. I was told that some in the Korean government offices even suspected that I had intentionally misled them into believing that UNDP would provide funding even though UNDP never had such a plan.

The second source of difficulties for IVI at this time was within the Korean government. When the Korea IVI Organizing Committee first sought to seek the government's commitment for the project, many believed that the most natural government partner would be the Ministry of Science and Technology (MOST). The Minister of MOST, Dr. Si-joong Kim, said his ministry wanted to be the supporting government partner for IVI. But he proposed to place the IVI headquarters in Daedeok Science City, which the Korean Government was developing as a major center for science and technology as a part of an ambitious national program, for which MOST was the lead agency. The Science City was located near Daejeon, a provincial city located more than 2 hours by car from Seoul, and equally distant from the main international airport in Gimpo. The Science City was still in its early stage of development, so it did not have the physical and social infrastructure to support an international organization. When I heard about the MOST plan, I told the Korea Committee members that in my view Daejeon would not be a competitive candidate for IVI, especially in comparison to other more cosmopolitan and well-developed cities such as Hong Kong, Singapore and Bangkok.

Seoul National University offered a winning alternative to the Korean Committee. Prof. Chong-un Kim, the President of SNU, proposed that IVI headquarters be located in the new research park being developed by the university on its campus. The SNU proposal would solve the problems of the Science City plan. However, since SNU was a national university, IVI would come under the administrative purview of the Ministry of Education (MOE), rather than MOST. While MOST officials were mostly technocrats with outward-looking ideas and accustomed to dealing with international issues, the MOE officials were usually associated only with domestic programs. In fact, many of my friends in Korean academia told me they considered MOE the most conservative and inward-looking branch of the Korean government. Prof. Cho, who had himself served as the Minister of MOE several years earlier, shared this view. It would have been better for IVI if MOST had been the official partner in the Korean government, instead of MOE. I was concerned that many bureaucratic issues would test the flexibility of the MOE officials.

Indeed, soon after we arrived in Seoul, we found that the mid-level officials of MOE were rather reluctant partners for us. UNDP's invitation document had stated that the host country would be expected to pay for 30% of the annual operating cost of IVI. Due to its budget situation mentioned above, the funding contribution from UNDP to IVI at this time was limited to paying the personnel costs of the UNDP team. The budget contribution from UNDP was certainly not insignificant, but with no other meaningful funding from foreign sources forthcoming yet, the size of the Korean government contribution for IVI quickly became a key issue. The Korean government position, as represented by MOE, was that Korea should pay for only 30% of the operating budget of IVI.

IVI was the first-ever international organization to be headquartered in Korea. As an international organization, certain official matters concerning IVI/UNDP were also the purview of the Ministry of Foreign Affairs and Trade (MOFAT). In October 1995, Korean President Kim Young-sam said in his speech at the United Nations General Assembly that Korea was hosting the International Vaccine Institute because Korea wanted to contribute to the health of the children of the developing countries through helping to develop better and more affordable vaccines. But the public policy announcement by the President and senior officials that Korea strongly supported the IVI initiative took time to filter down to the mid-level officials who actually controlled the budgets. The desk officers at the Ministry of Foreign Affairs and Trade and the Ministry of Education gave IVI staff grief whenever the issues of Korean government's contribution were on the table. In 1996, the problems reached a point where some of the officials at times tried to avoid meeting with me face to face.

Finalizing the plan for the construction of IVI building was also being postponed, year by year. The original plan had called for the completion of the building by the end of 1998, but the government did not commit the funds for the project and the delay added to the sense of uncertainty. Though not known to us at that time, the "Asian financial crisis" of 1997-1998 was already looming, and the general financial situation of the country was probably not favorable for such a large project.

Prof. Cho and Prof. Park were acutely aware of the difficulties that IVI was facing. Prof. Cho personally intervened, and devoted his time and effort to work with the government offices, using his personal influence gained from his long service as an academic leader and as a cabinet minister. Eventually, due to the mediation by Prof. Cho and his group of supporters, the Korean Government agreed to pay for the short-fall in basic maintenance cost of IVI for the first four years.

To have a more effective mechanism to work with the government offices and with other potential supporters, the *Korea Support Committee for IVI* was organized under the leadership of Prof. Cho. Many key members of the Korea IVI Organizing Committee, which had successfully brought IVI to Korea, became members of the Support Committee. IVI now had a reliable and trusted group to turn to. Even today in 2023, after more than 25 years, the Korea support Committee is still very active, and functions as the main non-governmental fund-raising channel that provides financial support for outreach programs for IVI, such as cholera and polio vaccination campaigns in Asian and African countries.

Since IVI could not begin laboratory-based research yet, we focused on training and technology assistance programs and network building. The first meeting of the collaborating partner institutions of Asia-Pacific countries was held at the Institute in Seoul in May 1995, with participants from China, Korea, Thailand, Australia, India and the Philippines. This group evolved into the more-formalized *Network Coordinating Group*, which was convened in Seoul in November 1996, to coincide with the IVI-sponsored *Symposium on Vaccines for the 21st*

Century, as part of the Pacific-Rim Biotechnology Conference held in Seoul. UNDP organized and chaired the *Institute Support Council*, into which we hoped to bring the major multinational vaccine companies and other NGO groups active in international health.

IVI staff scientists C.K. Lee and Stanford Lee started to give training workshops and lecture series on GMP production and quality control at the Korean National Institute of Health in Seoul and at the National Institute of Hygiene and Epidemiology in Hanoi. The first training course on GMP manufacturing was given in Bandung, Indonesia, in July 1996. Technical assistance and lecture series gradually expanded to Singapore, China, Thailand, Bulgaria and Iran. IVI staff also assisted in WHO inspections of vaccine production facilities in Korea, Brazil, Denmark and Bulgaria.

The first major opportunity to demonstrate the unique advantages that IVI offers as an international public research organization came to us in 1996. It was related to the introduction of a new vaccine against *Haemophilus Influenzae* type b (Hib) infection. Hib is a highly invasive disease of the very young children, with very high mortality and morbidity. Several multinational vaccine companies had each developed a new Hib conjugate vaccine, made by chemically linking a highly immunogenic synthetic protein component to the target antigen of the pathogen. The new vaccine was a major scientific breakthrough for public health, because it was clear that the disease burden for Hib dropped precipitously in the Western countries where the vaccine was first introduced. But the new vaccine was expensive to produce, and carried a correspondingly high price tag. For the vaccine producers, the new vaccine was a major source of new revenue.

It was reasonable to assume that the new Hib vaccine may have a similarly important public health impact in Asian countries, and therefore the vaccine producers could hope to see a large demand for the new vaccine there, greatly enlarging the vaccine market. However, reliable data regarding the Hib disease burden in any of the major countries in Asia were not available.

I reasoned that IVI would be the ideal institution to conduct a well-designed multi-center study to determine the Hib disease burden in Asian countries. Such a study would show whether the new Hib vaccine should be introduced in Asian countries, perhaps with public funding. Reliable data on disease burden was essential for the vaccine manufacturers because only population data would justify the vaccine use. At the same time, it would be more effective and much less costly for IVI, a public institution with its regional network in place, to coordinate such a study, rather than for each of the vaccine companies to do it on their own.

My plan was to organize a scientifically rigorous study, to be carried out with our partner countries under the supervision of internationally recognized experts. I felt there was a high likelihood that the major multinational vaccine companies will agree to jointly fund it. This would be a true public-private sector cooperative study to address an important public health issue for children in developing countries, as well as to respond to the needs of the vaccine companies.

In December 1996, I met with Prof. Joel Ward of UCLA Harbor Medical School, and asked him to direct a multi-country population-based study to determine the Hib disease burden in selected Asian countries. I told him that IVI will sponsor and coordinate the project. Joel Ward was Director of the UCLA Center for Vaccine Development, and was regarded as the foremost expert of the field. He accepted my invitation, and IVI's first large-scale scientific project was born.

Joel Ward and I were able to persuade several outstanding investigators to become members of the Study Committee: Prof. Xu Zhi-yi of Shanghai First Medical School of China, Prof. Ron Dagan of Ben Gurion University of Israel, and Prof.

Jung-soo Kim of Jeonju National University Medical School of Korea. (Prof. Xu later moved to IVI permanently as Senior Scientist.) Based on careful evaluations by the Study Committee of the local medical infrastructure and laboratory capabilities required for the study, 3 locations were selected as study sites: Hanoi, Vietnam; Jeonju City, Korea; and Guangxi, China. Two IVI scientists, Hai-Feng Huang and Joo-Yeon Kim were members of the study coordination team.

I visited each of the five major multinational vaccine companies to request their joint funding for the Hib study: Merck Vaccines in New Jersey; Wyeth-Lederle Vaccines and Pediatrics in New York; Chiron Corp. in California; Pasteur Merieux Vaccines in Paris, and SmithKline Beecham in Belgium. These five companies together were practically the core of the world's vaccine industry. It took several meetings at some of the companies, but eventually, all of them agreed to jointly support the IVI study. The fact that Joel Ward was the study director and that the field sites would be supervised by a distinguished team of outside experts provided the assurance that the study will be carried out according to internationally acceptable protocols. Also helpful, I believe, was that at each of these companies I had personal friends or scientist-executives who were sympathetic to the basic missions of IVI: Thomas

Vernon and Adel Mahmoud at Merck, Ron Saldarini at Wyeth-Lederle, William Rutter at Chiron, Stanley Plotkin at Pasteur-Merieux, and Francis Andre at SmithKline Beecham.

For these major companies to jointly fund a single multinational project was truly unprecedented. It was of course possible because IVI offered the most effective and scientifically credible way to conduct the study. At the same time, the joint funding of the Hib project signified that the global vaccine industry now accepted IVI as a true collaborator. The Hib study was formally launched in May 1997.

With the help of many dedicated supporters, both in Korea and in other countries, and with the continuing oversight provided by UNDP, IVI survived the myriad initial difficulties of the first years, and slowly gained international recognition as a new center of vaccine research for the developing world. The considerable challenges and uncertainties that IVI faced in the initial years of its operation were reported in a sympathetic full-page news article by Dennis Normile in the journal *Science*, under the title *"Vaccine Development: Korean Institute Ponders Role In Global Eradication Efforts"* (*Science*, December 6, 1996). It was the first substantial report about IVI by a major international scientific journal. The full report is reproduced in Appendix 2.

By late 1997, as technical assistance programs and

field studies steadily expanded, additional scientific and administrative staff joined IVI. In addition to the first three members who arrived in Seoul in January 1995 -- Gurinder Shahi (Singapore), Richard Mahoney (USA) and myself (USA) -- the staff now included Sae-joong Kim from Korea (Chief Administrative Officer); Sun-young Min from Korea (Executive Secretary); Prof. Xu Zhi-yi from China (Senior Scientist); Michael Klass from Sweden (Chief Financial Officer); Chung Keel Lee from USA/Korea (Chief Officer for Technical Cooperation); Soo-Young Stanford Lee from USA (Senior Technical Officer); Joo-Yeon Kim from Korea (Research Program Manager); Hai-Feng Huang from USA (Senior Scientist); Paul Kilgore from USA (Research Scientist); Vinay Gupta from India (Computer Systems Officer); Andree de Manuel from Spain (Information Officer), Hyi-sung Kim from Korea (Accounting Manager), and Eunyoung Kim and Kyung-hee Oh from Korea (Administrative Assistants).

9

Formal Birth of the International Vaccine Institute, 1997

On October 28, 1996, IVI passed the first milestone to becoming a free-standing organization when the Establishment Agreement was opened for signatures for member states at the United Nations in New York. The Secretary General of United Nations was the depositary. A Separation Agreement, officially establishing IVI as an autonomous institution independent of UNDP, was also signed by the UNDP Administrator James Speth and the Korean Ambassador to the United Nations Park Soo-gil.

The last formal milestone was reached on May 29, 1997, when the ratification documents for the Establishment Agreement were deposited with the Secretary General of the United Nations by 3 signatory states and WHO, the minimum number needed to put the Agreement into effect. IVI was now an independent international organization under the Vienna Convention. It was no longer an UNDP-managed organization. The long journey that began in October 1992 finally reached its destination.

We prepared for a celebration to mark the official birth of the International Vaccine Institute. The institute still faced many challenges regarding its financial sustainability, defining the nature and scope of its scientific programs, strengthening partnerships with other institutions in Korea and in other Asian countries, and resolving operating relationship with WHO and the Korean government. But the goal of establishing the new institute was now achieved, and we felt we deserved to celebrate.

On October 9, 1997, the establishment of the International Vaccine Institute was officially celebrated with a commemorative ceremony at the auditorium of Seoul National University Museum. Members of the Board of Trustees of IVI and diplomatic representatives from the Network Coordinating Group countries and UNDP attended the event. Many guests from Seoul National University, the Korean government and Korean academic circles also joined the celebration. Prof. Jeongbin Yim, who had offered his institute as temporary office for the arriving UNDP team in 1995, served as the Master of Ceremony.

As the director of the IVI Project, I delivered a short speech describing the history of the institute's birth and its mission. The speech is reproduced here, as it summed up the history of IVI and its mission as I viewed them at that time.

The Mission of the International Vaccine Institute

Seung-il Shin
Project Director for IVI
United Nations Development Programme
October 9, 1997

Minister (of Education of Korea) M.H. Lee, President (of SNU) J.W. Sonu, Your Excellencies, Members of the Board of Trustees, Distinguished Guests, Dear Colleagues and Friends, and Ladies and Gentlemen:

Today marks the official birth of a new institution, which in time will make great contributions towards better health of people everywhere, especially for the children in poorer countries. This week also marks the fifth anniversary of the start of the UNDP project which has now led to the creation of the International Vaccine Institute. As one member of the UNDP team that has coordinated the effort, I am particularly honored to be here today to celebrate the formal and successful end of that project.

The vision that guides the Institute has been embodied in the Preamble to the Constitution, which also defines its mission. To promote the development and introduction of new vaccines, the Institute will carry out research, training, technical assistance, service provision and information dissemination.

Under the guidance of the Board and with the advice and input from many experts from all over the word, the Institute is developing a Strategic Plan to set the future course of its scientific activities. When finalized, it will produce the basic guiding principles for the work of the Institute for the next several years. As a young institution, we must begin with highly focused programs. We have already initiated a multi-county epidemiologic research project on invasive bacterial disease in Asian children, with funding from the world's leading vaccine companies. Other programs on economic and policy analysis of vaccine introduction and use will begin soon.

As the Institute gains scientific and human resources, and

especially after the laboratory facilities become available in the year 2000, it will be able to engage in a broader spectrum of activities including laboratory-based research. The Institute will work in close partnership with other international organizations, particularly the World Health Organization, and with scientific institutions in Asia and elsewhere, and with industry.

Throughout the UNDP feasibility study for the Institute, I was impressed by the strength of the interest and support that many Asian countries offered to the idea of an international center for vaccine sciences designed to serve the needs of developing countries. There was also a strong consensus that the institute should be established as an independent international organization, in order to be free from national and institutional politics, dedicated solely to scientific work.

At the same time, by catalyzing the mobilization of intellectual talents, financial resources and policy commitment of the newly empowered Asian countries and by working together for a shared global cause, the Institute should bring in a new era of international collaboration among the countries of Asia, and between Asia and the rest of the world. I am delighted to note that these principles are now firmly imbedded in the Establishment Agreement and the Constitution of the Institute.

It has been a rare privilege for me to be a part of the international effort to establish the International Vaccine Institute. Such an undertaking of course would not have been possible without the support of many dedicated people, from UNDP, the Government of Korea, Seoul National University, and friends from many countries, as well as my colleagues at the IVI Office in Seoul. To all of them I am deeply grateful.

Thank you very much for your attention.

Selected photographs of the commemorative events of October 1997, showing members of the Board of Trustees and IVI staff members, are attached at the end of this history.

Our long-term optimism for the new institute was of course tempered by financial and political uncertainties that remained unresolved. Our hopes and concerns at this transition were accurately reflected in a news article in the journal *Nature*, dispatched from Seoul under the title "*Vaccine institute treads out a wary path*" (*Nature*, Oct 16, 1997). In an accompanying editorial commentary, *Vaccines at risk: An imaginative attempt to tap Asian resources for the benefit of the developing world deserves more support, Nature* urged the developed world to "chip in with support."

Below are excerpts from the *Nature* news article and commentary.

NEWS: Vaccine institute treads out a wary path
Nature Vol 389, p. 655 (16 October 1997)
By David Swinbanks

SEOUL: The world's first institution devoted to the research and development of vaccines for developing countries was formally established last week in Seoul, South Korea, when the United Nations Development Programme (UNDP) handed over the governance of the nascent International Vaccine Institute (IVI) to an independent board of trustees composed of eminent scientists and health officials.

But the new institute, which will be built in a science park on the campus of Seoul National University and is currently in temporary offices at the university, faces a difficult future as it defines a role for itself in the complex political arenas of world health care.

The institute was proposed by the UNDP in 1992. It comes under the umbrella of the Children's Vaccine Initiative, a broad coalition of organizations from public, non-government and private sectors with a secretariat at the World Health Organization (WHO) in Geneva that seeks to protect the

---

world's children against infectious diseases.

The Asian institute is the brainchild of Seung-il Shin, an American of Korean descent. Shin played a key role in the late 1980's in introducing affordable hepatitis B vaccines to the developing world, and felt the rapidly developing economies of Asia could provide a new source of funds and a means for developing affordable vaccines for the developing world

Seoul was chosen to host the IVI in 1994 after bids from several Asian countries ended in a three-way run-off between South Korea, China and Thailand. But the international agreement to establish the institute took effect only in May (1997), after three signatory countries ratified the agreement.

The institute also had to overcome initial opposition from WHO, which saw it as impinging on its own Western Pacific Regional Office in Manila in the Philippines. Some US vaccine manufacturer were also apparently concerned that it might become a commercial competitor.

WHO's opposition abated after three of its officials were appointed to the 16-member board of trustees, including the head of the Western Pacific Regional Office. Richard Mahoney, director of institutional development of the institute argues that the institute is a "new resource that will attract new funds", as witnessed by the substantial contributions from the Korean government and the contributions from Western vaccine manufacturers...

Vaccines at risk: An imaginative attempt to tap Asian resources for the benefit of the developing world deserves more support
Commentary, *Nature* Vol. 389, p. 642 (16 Oct 1997)

Sadly, the developed world seems to be so unwilling to donate money to the prevention of disease in the developing work that backers of two worthy initiatives to develop vaccine for children may end up fighting over the same small pot of funds (see page 655). Promoters of the International Vaccine Institute in Seoul, who realized that there is new money in the (until recently) booming economies of Asia, have succeeded in winning substantial Korean government support for the institute's construction and 30 percent of its running costs. But some officials of the World Health Organization (WHO) are concerned that, in pursuing the remaining $10 million or so a year that will be needed to run it, the new institute may eat into the small $30-million pie of the WHO-backed Children's Vaccine Initiative which is itself responsible for the institute.

Both initiatives seek to provide vaccines to prevent disease in children with, in the case of the Seoul institute, a focus on the developing world, and in particular Asia. Is the world so poor it cannot afford the $40 million required to support both initiatives? There were some understandable – but misguided

– concerns that the institute was a Korean attempt to tap into Western vaccine technology for the benefit of Korean industry. But the establishment of a distinguished international board of trustees to oversee the institute and a clear statement that it will not engage in the sale of vaccines should dispel such fears. The developed world, including Japan and the United States, should chip in with support.

With the opening of IVI, my official mission with UNDP finally came to an end. Before I could return to my home and family in the United States, however, we needed to recruit a new director of the institute in its newly-gained status as an autonomous organization, independent of UNDP oversight. But the recruiting process proved to be a lengthy one involving false starts and reruns, and only in July of 1999 was the new director, John Clemens, appointed by the Board. John was a well-known clinical epidemiologist at the National Cancer Institute of the U.S. NIH, who had formerly worked at the ICDDRB in Dhaka, Bangladesh.

Meanwhile, IVI continued to expand its programs and to provide technical services. The Hib study was carried out excellently on schedule. In early 1998, *Nature Medicine* published a special supplement on vaccines, and I was invited to contribute a paper on the global situation in vaccine research and its implications for the developing world. It was evident that IVI was now receiving recognition by the international scientific community as an important player on the global scene. The invitation gave me an excellent platform to call for greater international support for IVI's work. My contribution (*Nature Medicine*, Vol 4, May 1998) is reproduced in Appendix 3.

By June 1998, 32 countries and WHO signed the IVI Agreement. The government of Korea finally approved the funding for the construction of the headquarters building. We formed a committee to review and approve the architectural design for the building. The groundbreaking ceremony was held on August 18, 1999, with the Prime Minister of Korea, Jong-pil Kim, in attendance. It was the last official IVI function that I participated as the Director of the IVI Project and as a UNDP officer.

10

End of the Mission

In October of 1999, I moved to San Francisco to join a dedicated group of scientists at VaxGen, a small group that was conducting a multinational clinical trial of a promising new vaccine for AIDS, called *AIDSVAX*. Genentech had developed the vaccine based on a recombinant viral surface glycoprotein of HIV, the virus that causes AIDS. VaxGen was carrying out the efficacy trials in the U.S., The Netherlands, Thailand and Puerto Rico. It was the only AIDS vaccine that was in advanced human clinical trials at that time. The stakes were extremely high, because the AIDS epidemic was the most pressing global health issue of the day, causing the deaths of millions of people every year, most of them in the world's poorest populations. My first responsibility at VaxGen was to develop a plan for the production and global distribution of the vaccine, if and when the clinical trials proved the vaccine's efficacy. The search for a manufacturing facility that could produce at least 100 million doses of *AIDSVAX* per year showed me that no existing vaccine company in the world had such a capacity. My answer was to create a new company in Asia to take on that role. A biopharmaceuticals manufacturing company called Celltrion, launched in 2002 in the Incheon Free Economic Zone in Korea, was the result. But this is another story.

As I left IVI to return to America in 1999, I felt great satisfaction that the original mission to turn the idealistic dream of IVI into reality was finally achieved. But I was also filled with a sense of regret that IVI was not on a firmer financial ground, and that I did not see IVI in its own new building.

In the following years, however, I was able to observe happily from a distance as IVI established itself as a major international center for vaccine sciences for the world's poor. Major grants from the newly-founded Bill and Melinda Gates Foundation were an important boost for IVI. It was also reported that the Korean government's financial support greatly increased and stabilized, and that new funding from other sources also began to arrive.

In 2003, IVI finally moved into the newly-built headquarters building, a beautiful and imposing modern structure on the hillside of the Gwanak Mountain in Seoul.

Coda: IVI, Twenty-Five Years Later

On October 20, 2022, IVI marked its 25th Anniversary with a gala celebration at the Lotte Hotel in downtown Seoul. It was a happy, sparkling event on a beautiful autumn day. Under the leadership of Director-General Jerome Kim, IVI has greatly enhanced its reputation as the indisputable global center of vaccine sciences serving the needs of the developing world. According to the *25th Year Impact Report* issued by IVI, the list of its accomplishments over the past quarter century was truly impressive. The report said that more than 3,000 vaccine professionals from all corners of the world received training through IVI's annual International Vaccinology Course. IVI was pivotal in the development of two major vaccines, an oral cholera vaccine and a typhoid conjugate vaccine, and was working on vaccines for 9 other infectious diseases. More than a million people have been vaccinated through IVI campaigns across Africa and Asia. A young Korean vaccine company, *Eubiologics*, produced and distributed more than 100 million doses of the cholera vaccine in endemic countries by early 2023. IVI now has 39 countries and WHO as signatory members, with research collaborations going on in 44 countries. In addition to the headquarters in Seoul, IVI has opened European Regional Office in Stockholm, a country office in Vienna, and collaborating centers in Ghana, Madagascar and Ethiopia.

On this celebratory occasion, IVI awarded the Founder's Medal to Wan-kyoo Cho, Sang-dai Park, Barry Bloom and me. It was the first time in nearly 25 years that the four of us got together again in one place, and we were able to look back at the shared years of our lives devoted to make IVI in Seoul a success story. Many old memories flooded back to me.

Acknowledgements

I cannot close this story without remembering, and thanking, the many colleagues and friends who worked with me in the collective journey for the International Vaccine Institute. Foremost is Frank Hartvelt of UNDP, who first saw my dream for IVI and made it a mission of his own. Frank stood steadfast and pushed us forward, even when the vision of IVI seemed almost impossible to achieve. I was especially fortunate also to have had the trust and support of Gurinder Shahi and Richard Mahoney, two close colleagues who helped me when I needed their help most. As I have recounted above, it was the efforts of Prof. Wan-kyoo Cho and Prof. Sang-dai Park, whose humanitarian idealism and patriotic devotion to promoting

Korean science made IVI the Korean success story it is today.

To my former colleagues at IVI and at UNDP, and to the many colleagues and friends in Korea and other countries who shared with me the vision that IVI stood for, I am deeply grateful. In Appendix 1, I have collected photographs of the key events in the early history of IVI, and of the many partners and colleagues who joined forces to make the birth of the new institute possible.

Appendices

IVI's Early History in Photos

Site Selection Committee team in Korea (Chairperson M Catley-Carlson is at front center) (May 1994)

IVI Delegation at UN (left to right, G Shahi, T Rothermel, a UN officer, B Bloom, F Hartvelt, W Cho, S Park, S Shin, R Mahoney) (1995)

IVI celebrates formal Establishment (October 1997)

First Meeting of Governing Board of Trustees (October 1997)

IVI staff with Board members (October 1997)

Korean President Kim Young-sam (front row, fourth from left), meeting with IVI senior staff and Board members at the Blue House (October 1997)

Groundbreaking Ceremony for IVI Headquarters Building at Seoul National University (August 1999)

Groundbreaking Ceremony (August, 1999)

Adolfo Martinez-Palomo, John LaMontagne and V. Ramalingaswamy (1997)

Gus Nossal and Seung-il Shin (1997)

Jeongbin Yim and Gurinder Shahi (1997)

Jeongbin Yim and Seung-il Shin (1997)

Richard Mahoney and Seung-il Shin

Michael Klass and Joo-yeon Kim (1997)

Moving into new IVI office space (1999)

SNU professsors visit IVI office. (Left to right), Sang-dai Park, Seung-il Shin, Kijoon Lee (President of SNU), Wan-kyoo Cho and Sang-cheol Park (1999)

IVI staff and friends on my departure day (September 1999)

Sang-dai Park, Wan-kyoo Cho and Richard Mahoney at the IVI building (2009)

Founders Medal Awardees (Left to right) George Bickerstaff (Board Chair, IVI), Seung-il Shin, Barry Bloom, Wan-kyoo Cho, Sang-dai Park, Jerome Kim (DG, IVI) (October 2022)

Prof. Ruth Arnon
(Israel)

Prof. Chen
Chunming (China)

Prof. Wan-kyoo Cho
(Korea)

Drs. Darodjatun
(Indonesia)

Prof. Demissie Habte
(Ethiopia)

Dr. John LaMontagne
(USA)

Dr. Richard Mahoney
(UNDP/USA)

Dr. Adolfo Martinez-
Palomo (Mexico)

Prof. Sir Gustav
Nossal (Australia)

Dr. Akira Oya
(Japan)

Dr. Lars Pallesen
(Denmark)

Prof. Sang-dai Park
(Korea)

Prof. V. Ramalingaswamy
(India)

Dr. Geoffrey Schild
(UK)

Prof. Seung-il Shin
(UNDP/USA)

Key collaborators not present here are Prof. Barry Bloom (USA), Frank Hartvelt (UNDP/Netherlands), Timothy Rothermel (UNDP/ USA), Prof. Philip Russell (USA) and Dr. Gurinder Shahi (UNDP/ Singapore).

Appendix 2.

Science, Vol 274, p. 1607 (6 December 1996)

Vaccine Development: Korean Institute Ponders Role In Global Eradication Efforts

Dennis Normile

Seoul – The eradication of small pox in 1977 proved that industrialized nations can mount a successful global immunization program with an arsenal that includes research, training, public education, and distribution of the right vaccine. Polio and measles are next on a list of possible targets. In most cases, these vaccines had initially been developed for disease that affected the industrialized world. Yet health officials and scientists say that progress against infectious diseases that primarily affect developing nations, such as malaria and tuberculosis, will require greater attention than is now being given to them by the industrialized world.

It is hoped that a new International Vaccine Institute (IVI) taking shape here will redress this lack of attention by marshaling the talents and resources of the developing countries themselves. However, its success depends upon making the most of a slim budget, and avoiding duplicating existing vaccine development efforts.

The idea for the institute grows out of the Children's Vaccine Initiative, begun in 1991 by a host of international organizations to provide affordable vaccines for children. On 28 October, the IVI passed a major milestone with a ceremony at the United Nations that established it as an autonomous, not-for-profit institute. Last month, to celebrate the occasion, the institute sponsored a symposium on *Vaccines for the 21st Century* as part of the 5th Pacific Rim Biotechnology Conference in Seoul. Meetings of an IVI network coordinating group and the IVI advisory board turned the week into a brainstorming session on how IVI can make its greatest contribution.

IVI's major goal is to expand the roster of existing vaccines to meet the public health needs of the developing world. "Some very important diseases against which vaccines need to be developed are not being developing because [they] do not have obvious commercial potential," says Seung-il Shin, a Korean-born biochemist and former U.S. academic researcher and entrepreneur who is project leader for the institute as it searches for a director. Richard Mahoney, a public-health expert and IVI's director of institutional development, says that possible targets include malaria, tuberculosis, Japanese encephalitis, and *Haemophilus influenzae B.*

In 1992, Shin joined the United Nations Development Program (UNDP) to work on a series of studies that led to IVI and the selection of Seoul as its home. The South Korean government promised up to $50 million to build and equip a facility for 40 to 50 researchers, including those who might set up satellite laboratories, as part of an overall staff of 200 on the campus of Seoul National University. Construction is expected to be finished in late 1999.

The government has also agreed to pay 30% of an estimated $15 million a year in operating costs. Shin hopes the rest will come from international donor agencies and other East Asian countries, supplemented by contract research and fees from educational and training programs in vaccine production and delivery. "Obviously, [funding is] a major challenge," he says.

The institute's constitution describes an ambitious agenda that includes research and development, education and training programs, technical cooperation to boost the research and production capabilities of developing nations, and disseminating information through publications and conferences. Adolfo Martinez-Palomo, director-general of Mexico's Center for Research and Advanced Studies and a member of the IVI board, believes that developing countries need "to establish and strengthen an adequate health research base." He points to the recent discovery that what had long been considered two forms of amebiasis, a parasitic disease, are actually two different infections, one relatively harmless and the other a major killer in developing countries.

Some public health administrators feel IVI is in danger of stretching its resources too thinly. "My major problems in selling IVI within WHO and to other countries is its all-inclusive constitution," says Jong-Wook Lee, a Korean-born physician who heads WHO's Global Program on Vaccines and Immunization and is also executive secretary of the Children's Vaccine Initiative. Support from countries that are already contributing to vaccine-related work at WHO and UNICEF is not likely, he says, unless IVI can make clear its unique role. "In my views, IVI should focus on vaccine R&D."

But even in R&D, IVI will have to focus its efforts. Trying to do everything on its own "would be not only foolish but impossible," Shin agrees. He says IVI doesn't want to duplicate basic work done elsewhere, nor is there any point in competing with pharmaceutical companies.

Joh LaMontagne, a U.S. National Institute of Allergy and Infectious Diseases official who served on the IVI site-selection committee, says a recent study by his institute found that there were 250 basic research candidates that could possibly be turned into vaccines. But only a handful of these are likely to go further along the development pipeline. "The major challenge facing all of us is the problem of translating these discoveries of basic science into reality," he says.

IVI's Mahony says the experience of the International Task Force on Hepatitis B Immunization, which he chairs, could provide a model of how IVI could relieve some of the bottlenecks in the vaccine development process. With a budget of only $7 million over 9 years, the task force worked jointly with the industry to conduct marketing studies, sponsor model effectiveness trials, and help companies clear regulatory hurdles in particular counties. As a result, he says, what in 1986 was a very expensive vaccine given only to high-risk groups is now plentiful, relatively cheap, and given routinely to 10% of the world's newborn infants. That success story is well known to Shin, who before joining UNDP was a partner in a New Jersey biotechnology firm working on one version of the vaccine.

Philip Russell, a public health expert from Johns Hopkins

University and a member of IVI's networking group, says that in the developed world, public health authorities often cooperate with the private sector to bring to market vaccines or drugs not commercially attractive enough for private companies to tackle on their own. "There is nobody playing that [public sector] role in Asia," he says. IVI's challenge, he adds, is to fill this gap.

Appendix 3.

Nature Medicine Vol 4, p. 503-505 (May 1998)
Vaccine Supplement

The Global Vaccine Enterprise: A Developing World Perspective

The industrialization of the vaccine enterprise has implications for the supply of vaccines to the developing world

Seung-il Shin

International Vaccine Institute
Seoul National University Campus
Shillim-Dong, Kwanak-Ku, Seoul, Korea

THE MOST IMPORTANT factor driving the transformation of the vaccine enterprise (which encompasses the development, clinical testing, production, licensure and distribution of vaccines) is the increasingly complex scientific and technological base that is required to develop and manufacture the newest generation of vaccines. The traditional (and highly successful) vaccines, such as those against small pox, diphtheria, tetanus, pertussis and tuberculosis, were based on the pioneering work of Jenner and Pasteur. The pathogen was grown in quantity in a simple facility, purified in a few steps, killed with an inactivating agent (where appropriate), and blended into the final product. Within a few decades of Pasteur's death, his disciples established Pasteur Institutes or similar public institutions in many parts of the world that produced vaccines as a public service (1).

In contrast, today's new vaccines are 'high tech' products that require expertise in multiple scientific disciplines, large numbers of skilled staff, and costly advance investment in research and manufacturing facilities. New generation vaccines, such as genetically engineered subunit vaccines against hepatitis B virus, cell-free vaccines against whooping cough (pertussis), and the protein-polysaccharide conjugate vaccines against invasive bacterial diseases, for example those caused by *Haemophilus influenzae* type b (a major cause of bacterial meningitis in small children) and *Streptococcus pneumonia*, bear little resemblance to the traditional vaccines in the way that they are produced.

The second factor driving the transformation of the vaccine enterprise is the changing nature of technology ownership. Even though the basic research supporting development of vaccines is conducted at public and academic research centers supported by public funds, vaccine development has become primarily the purview of large industrial laboratories, often augmented in key segments by specialized biotechnology companies funded by venture capital. Thus, most key technologies for future vaccines will be developed and owned by companies that will diligently protect their new inventions through internationally enforced patents. In Pasteur's day, and even as recently as forty years ago when the polio vaccines were first developed, most of the new technologies needed to manufacture vaccines were owned by the public. The scientists and organizations that developed them often assisted and funded the technology transfer to institutions in developing countries. It is improbable that the developing world will have such easy access to key vaccine technologies in the future.

The third factor is the globalization of international commerce. In order to compete successfully, vaccine companies have been consolidating on an ever-larger scale. The global vaccine industry in 1998 is thus dominated by a small number of large multinational companies, instead of the smaller, publicly owned and public-spirited national vaccine production centers that until recently were the norm. Consequently, some of the key decisions regarding which vaccines to develop and how to distribute (market) them are no longer made by scientists and public health officials but by business executives in board rooms who must answer to their stockholders.

Finally, the increasingly stringent international product safety standards required of vaccines have particular implications for the supply of vaccines to developing countries. Heightened international standards for vaccines may be only a reflection of the larger trend of higher safety standards for consumer products in general, but the issue of product safety for vaccines is much more complex than for other pharmaceuticals

Traditionally, the protection of the public against major infectious diseases was the primary goal of public vaccination efforts. The provision of vaccines was often viewed by national governments as an essential public service. In an age of widespread public concerns about paralytic polio, for instance, the public was more willing to accept a vaccine that could save a million lives even if it might cause inadvertent but unavoidable harm to an unfortunate few. However, when polio is no longer a threat to most people because of the very success of polio vaccination, even a few cases of vaccine-associated polio may be deemed unacceptable.

But in developing countries where polio is still a potential public health concern, protection of the general population has to take precedence over the avoidance of a few adverse reactions. In many industrialized countries, public pressure to minimize vaccination-associated risks as well as ever-increasing concerns over product liability and lawsuits are forcing the vaccine industry to adopt the maximum defensive policy, by opting for the highest possible level of product safety. Because the safety-related technology keeps improving, the product safety requirements concomitantly become more stringent. The added cost will eventually be passed on to consumers everywhere as higher vaccine prices.

Newly introduced international safety requirements have the effect of making older vaccines, produced according to previous standards, potentially obsolete. Several of the larger developing nations together produce as much as 65 percent of the world's total output of traditional childhood vaccines used in the Expanded Program in Immunization (an international effort to increase vaccination coverage for six major childhood diseases coordinated by the World Health Organization and UNICEF). But few of these vaccines are exported to other developing countries because of their inability to meet the current international safety standards (2). Meeting these requirements is not an easy task for most Third World vaccine producers. Their facilities were established many years ago and would require major new investments in personnel training and new construction, as well as the introduction of national safety control laboratories. Reaching an acceptable balance between the need to keep up with the evolving technologies in vaccine production and the need to make vaccines affordable to the greatest number of people will be a continuing policy challenge for developing nations.

The key concerns of the developing world in the areas of vaccine production and supply can be summed up as access, affordability, equity and national autonomy.

A highly effective and safe vaccine against hepatitis B was developed in the early 1980s, but its incorporation into immunization programs in developing countries was long prevented by its high cost. It took a major concerted effort by an international group of dedicated people to introduce the vaccine to several developing countries in Asia and Africa but the key element in its eventual success was a drastically lower price (3). However, more than 15 years after its development, the hepatitis B vaccine is not as broadly available today as it should be, and the price factor is often cited as the most important reason. Other vaccines of importance to developing countries, such as the new acellular pertussis vaccines, the conjugate vaccines against *Haemophilus influenza* type b and, we hope, an AIDS vaccine, will not be widely available for the world's poorer citizens unless they are affordable. A stark example of the national choices that developing countries are facing is the recent announcement that India will launch its own program to develop an AIDS vaccine (4).

New candidate vaccines against major infectious diseases will need to be evaluated among the populations of the developing world where they will be most used. This is particularly true for vaccines against HIV, malaria, tuberculosis, diarrheal diseases, and acute respiratory infections, which have global significance but predominantly affect the developing world. However, research involving human subjects raises complex ethical issues, as was highlighted recently by the heated debates regarding the testing of potential AIDS vaccines (5).

Many developing countries appear to hold the view that vaccines are an essential public commodity and that a degree of national autonomy in vaccine production capability must be maintained, even at considerable economic cost. There is often a sense that a populous sovereign state should not become overly dependent on foreign (particularly commercial) suppliers for something as critically important as vaccines. In fact, many countries, including China, India, Indonesia, Brazil, Cuba and Mexico, have recently strengthened vaccine development and production, even though it could be more economical to import some of these vaccines from a foreign supplier. Whether it is justified or not, it is unlikely that the desire for national autonomy will disappear soon.

To seek creative solutions to some of these problems, the United Nations Development Program, with the help of many other organizations and major financial support from the Government of Korea, has created the International Vaccine Institute (IVI). This research center will be headquartered in new buildings on the campus of Seoul National University in Korea and will provide assistance to individuals and institutions in the developing world, so that they may become active participants in the evolutionary process that is reshaping the vaccine enterprise. The constitution of IVI (www.ivi. org), embedded in a United Nations-sponsored international agreement, has been signed by more than 30 countries and the World Health Organization and was ratified in 1997.

Is protection from diseases that can be prevented by vaccination a universal human right? If the answer is even partially yes, we must look for ways to ensure that a reasonable minimum level of access, affordability and equity for essential vaccines is provided for all. The transformations now taking place at the global level are inevitable. So will be the demand to find realistic solutions to address the legitimate needs of developing nations. The solutions must be acceptable to both the public and the private sectors, and to both the developed and the developing worlds.

1. Plotkin, S.L. & Plotkin, S.A. A short history of vaccination. In *Vaccines*. (eds. Plotkin, S.A. & Mortimer, E.A.) pages 1-11 (W.B. Saunders Co., New York, 1994).

2. Shin, S. & Shahi, G.S. Vaccine production in developing countries. In *Vaccination and World Health*. (eds. Cutts, F.T. & Smith, P.G.) pages 39-60 (John Wiley and Sons, Chichester, 1995).

3. Muraskin, W. The war against hepatitis B: A history of the International Task Force on Hepatitis B Immunization. (University of Pennsylvania Press, Philadelphia, 1995).

4. Jayaraman, K.S. India to develop its own AIDS vaccine. *Nature Med.* 4, 7 (1998).

5. Bloom, B.R. The highest attainable standard: Vexed ethical issues in AIDS vaccines. *Science 279*, 186-188, 1998.

Seung-il Shin

Shin was born in South Korea and spent his childhood there, when the country was still overwhelmed by the devastations brought by the Korean War and its aftermath. He entered Seoul National University, but left the college without earning a degree, to work as a reporter for the English-language daily newspaper, *The Korean Republic* (since renamed *The Korea Herald*), where he reported on social and cultural issues.

He resumed academic carrier by entering Brandeis University in the United States with the help of the Lawrence Wien International Scholarship. From Brandeis, he received a bachelor's degree in chemistry and a doctorate in biochemistry. He continued his scientific work in Europe, first as Research Fellow at the Institute for Human Genetics at Leiden University in The Netherlands, and as Visiting Scientist at the National Institute for Medical Research in Mill Hill, London. He then moved to the newly-opened Basel Institute for Immunology in Basel, Switzerland, as a member of the founding group of scientists.

Shin returned to America, and served for 14 years as professor of genetics at Albert Einstein College of Medicine in New York. As the genetic engineering revolution ushered in a new era of improved vaccines and biotherapeutics, Shin left the academia to co-found and run a biotechnology company called Eugene Tech International in New Jersey. A key project of the company was the development of a low-cost hepatitis B vaccine in cooperation with Alfred Prince of the New York Blood Center. Shin became a leading advocate of making the new hepatitis B vaccine available to the developing world, but he was often blocked by the political and economic realities of the international market place. This experience prompted him to propose an international center for vaccine research, development and training, focused on the needs of the developing countries. The United Nations Development Programme adopted his proposal, and invited him to lead a study to test the feasibility of creating the International Vaccine Institute.

This memoir recounts the history of the early events that eventually led to the formal launch of the IVI in Seoul in 1997.

www.ingramcontent.com/pod-product-compliance
Lightning Source LLC
Chambersburg PA
CBHW071754150726
47998CB00005B/1929